the story of mathematics

To Marina &
in memory of Paul

the story of mathematics

Richard Mankiewicz

CASSELL PAPERBACKS

contents

foreword

I've finally realized that for a large part of my life I've been fighting to break down a short circuit in the collective mentality of my fellow citizens. The essence of that short circuit can be summed up as the equation *maths = school*. Mention mathematics to most people and the immediate response is something about their school experiences. Not so long ago there was 99 per cent probability that it would be 'I was never any good at maths when I was at school', said with a curious kind of chip-on-the- shoulder defiant pride. But by 1995, going to a party and announcing yourself to be a mathematician was quite likely to trigger a lengthy discussion of fractals, chaos theory and the Santa Fe Institute. By the end of the '90s the favoured topic was Fermat's last theorem. (If you have no idea what any of these are, then you desperately need to buy this book.) But even in the year 2000, the majority of people still associate maths with school, *and with nothing else*.

The 'school' bit I can live with. The 'nothing else' is awful. How can we combat this terrible, all-pervasive ignorance about one of the major forces behind the creation of the modern world, and one of the central strands of human intellectual activity?

Not all at once, certainly. There has been a distinct improvement in the public understanding of mathematics, and its acceptance as an entirely reasonable and normal human activity. My evidence is the increasing willingness of the news media to report advances in mathematics, or new applications of it – and not just hidden away in the science pages (remember the biscuit-dunking formula?). Then there's the remarkable popularity of mathematical books and magazine articles for non-specialists. Maths books have topped the non-fiction bestseller lists. Maths movies have won awards.

How has this come about? Not through any major international movement or government initiative. I write this during what UNESCO has declared to be World Mathematics Year 2000, and the response from the UK government has been to seize upon the *maths = school* equation and then do very little with it. No, the newfound discovery 'that maths is sexy, maths is the new rock 'n' roll' – I quote a major national newspaper – has come about because of the uncoordinated activities of large numbers of individuals. Each one has found his or her own way to take one aspect of mathematics and make it accessible to a wider audience.

And so, piece by piece, a new consensus on mathematics has been building, in which the subject is seen as a frontier area of scientific research, a central force in the growth of technology, and a civilizing influence on human culture. It has always been all these things, but now more people are noticing.

Richard Mankiewicz is one of these dedicated lone wolves. I first met him in Croydon at an exhibition of the work of the Dutch artist Maurits Escher. Escher was not mathematically trained, but his almost surreal works are dominated by intensely mathematical themes – tiling patterns, non-Euclidean geometry and a kind of philosophical visual pun at whose core is the purest mathematics.

Richard wasn't an organizer of that particular exhibition, but for years he's been putting boundless energy and enthusiasm into imaginative projects to bring mathematics to the people. And one of those is *The Story of Mathematics*, a book that he wrote, he tells us, because 'it didn't exist'.

Well, it does now. And here's what it tells you. Mathematics does not just consist of a few arithmetical tricks that you learn at school and then promptly forget when you become an adult. It has an unbroken history of involvement with the mainstream of human culture, a history that has been going for at least five thousand years. And unlike the arts, this is not just five thousand years of *x* influencing *y* in some minor way, but five thousand years of *y* building directly on the work of *x*. Mathematics has been a collective activity of a relatively small number of unusually talented individuals who have cut across spatial and temporal boundaries as if they did not exist, and together created one of the wonders of the world.

When I was at school I spent many hours scouring the local libraries for books about maths. Nobody had told me that teenagers didn't do that sort of thing, and if they had done, I wouldn't have taken any notice anyway, because I was already hooked. To be truthful, the range of books on offer wasn't fantastic. Back then, there were very few suitable books about maths, and I read them all. Among them were several 'histories of mathematics', notably the vivid (but often inaccurate) writings of Eric Temple Bell, like *Men of Mathematics* and *Development of Mathematics*. But there was nothing like *The Story of Mathematics*: visually stunning and aimed at the cultural core – the ongoing interaction between mathematical thinking and all other human activity.

Mathematics has played a key role in bringing us map-making, navigation, perspective in art, radio, television, and the telephone. Without it, airlines would be unable to run efficiently, satellite TV would have a tenth as many channels and the world's food supply would be unable to sustain the current population. I'm not saying that we owe these things to mathematics alone, but maths was an essential component. And I'm not saying that these things are necessarily *good*, but they're sure as heck *influential*.

This, then, is our subject: one of the longest, most brilliant threads in the tapestry of human history, and one that is most intimately entwined with the thread of human achievement. And *The Story of Mathematics* takes us through many of the core events in a simple and comprehensible manner, with stunning illustrations that prove you don't *have* to do mathematics on the cheap.

It's the sort of book I would have loved to have read when I was a teenager. But, to return to my opening theme, this is a book for everybody. Being an adult doesn't let you off the hook. Time you acquired some *real* culture.

IAN STEWART

preface

'And what is the use of a book,' thought Alice, 'without pictures or conversations?'

Lewis Carroll, *Alice's Adventures in Wonderland*, 1865

This book was created because it didn't exist. I was searching for a way of representing the history of mathematics in an accessible style. Rather than taking the reader through a sequence of 'great theorems', I wanted to illustrate how the mathematical sciences were intimately linked to the interests and aspirations of the civilizations in which they flourished. I thought this would be best achieved by combining the visual aspects of mathematics with the written comments of mathematicians themselves, all brought together with sketches of the historical periods in question and key developments in mathematical ideas. The demands of space and time mean that I cannot tell the whole history of mathematics. Instead, this book presents selected highlights of a story that ebbs and flows with the changing fortunes of the some of the world's great civilizations.

From its beginning, mathematics has been apparent in every facet of human activity. Trade, agriculture, religion, war – all have felt the influence of mathematics and all in turn have influenced the concerns of mathematicians. And yet its history remains largely hidden from our cultural gaze. But I would actually go so far as to say that the evolution of science, philosophy and mathematics, all related, is far more important to the history of humanity than a parade of rulers and a procession of wars. In a society of unparalleled scientific achievements, I hope this book will make a small contribution to a scientific culture.

Perhaps the sciences, and especially mathematics, have lacked the kind of high public profile that has been enjoyed by the arts, and thus have not engaged people's hearts and minds in the same way. Some cross-fertilization has already taken place, with concepts such as relativity, quantum mechanics and artificial intelligence and the incompleteness theorem all becoming part of the stock of contemporary ideas. But when mathematicians talk about the beauty of their subject, it is often viewed as an embarrassing vestigial emotion expressed by those who have spent too long in the rarefied atmosphere of an ivory tower. However, the use of computers has finally made the beauty of mathematics accessible to all.

Mathematics is not about impenetrable symbols. It is about ideas: ideas of space, of time, of numbers, of relationships. It is a science of quantitative relationships, whose growth in sophistication and subtlety has mirrored humanity's quest for knowledge. And all ideas are born of a vision. With the growth of computing power, it has been reborn as a visual science. The extraordinary structures such as are found in chaotic and complex systems cut through the forest of symbols and open up to everyone a direct view of the mathematicians' landscape. A new aesthetic is in the air, combining mathematical precision with artistic sensibility. A large part of this book is devoted to illustrating that such a blend has been ever-present in varying degrees. The two cultures have had a very long period of engagement, though they have never as yet made it to the altar.

chapter one

year zero

◄ A terracotta table of accounts in cuneiform script from *c.* 2400 BC.

Any book must have a first chapter with an opening sentence. History is not so neat and tidy, and the search for the first use of numbers is a journey into the misty origins of human life and civilization. Archaeologists and scholars strive to construct a meaningful mosaic of our prehistory from a handful of tiles. New discoveries are not merely additional pieces of the puzzle, but may radically alter our whole picture of the past and our relationship to it. We must bear this in mind as we look at some of the earliest testaments to mathematical activity, and then at the mathematical cultures of Mesopotamia and Egypt.

The earliest evidence of numerical recording was excavated in Swaziland, southern Africa, and consists of a baboon's fibula with 29 clearly visible notches, dating from about 35,000 BC. It resembles the 'calendar sticks' still used in Namibia to record the passage of time. Other bones from the neolithic period have also been found in Western Europe. A radius of a wolf that was found in the Czech Republic and dated to about 30,000 BC, is marked with 55 notches in two series of groups of five. This appears to be a tally, possibly a record of animals hunted. One of the most intriguing finds is the so-called Ishango Bone, discovered by the shores of Lake Edwards between Uganda and the Democratic Republic of Congo. Dated to about 20,000 BC, it seems to be more than a mere tally stick. Microscopic analysis has revealed further markings which suggest a link to the phases of the Moon. Given the importance of predicting the full Moon, possibly for religious reasons but certainly for practical ones of nocturnal visibility, it is not surprising that keeping track of the big clock in the sky should be a major concern of neolithic peoples. In fact, be it through astronomy, astrology or cosmology, the heavens have probably been the single biggest influence on the development of mathematics.

From Mesopotamia, the land between the Euphrates and Tigris rivers, we have written records going back to about 3500 BC. The region had a whole sequence of cultures ruling over it. The early Sumerians and Akkadians were followed by the iron-craftsmen, the Hittites, who succumbed to the fearsome Assyrians. The Chaldeans and their famous King Nebuchadnezzar followed, but were subsequently overthrown by the Persians, who in their turn were overrun by the armies of Alexander the Great. The centre of power shifted between the cities of Ur, Nineveh and Babylon. Our main sources of mathematics come from the Old Babylonian Empire (1900–1600 BC), which showed influence of both Sumerians and Akkadians, and the post-Alexandrian Seleucid dynasty of the fourth century BC, in which both Greek and Babylonian influences are apparent. Because of the key position of Babylon throughout this period, the mathematics is often referred to as Babylonian.

Our present, decimal number system is a place-value system with a base of 10 – in other words, ten units in one counting place are equivalent to one unit in the next-higher counting place, and the position of a digit in a number determines its value. The earliest writings we have show that the Babylonians employed a sexagesimal, or base 60, number system. It has survived to this day in our method of timekeeping. Thus, for

▲ A Babylonian mathematical tablet showing a multiplication table. Clay was plentiful in Mesopotamia and hand-held tablets were used by students as exercise books. As long as the clay remained moist a calculation could be erased and a new one started. Dried-up tablets were discarded, some being used in the foundations of buildings, to be discovered many centuries later.

example, the Babylonians would express the number 75 as '1,15', which is identical to our writing 75 minutes as 1 hour and 15 minutes. From around 2000 BC a place-value system emerged which used only two cuneiform symbols, T for 1 and < for 10, while retaining the sexagesimal base. Thus 75 would be written as T<TTTTT. The number system was greatly extended by having sexagesimal fractions, but there was no symbol for zero. A positional symbol did not come into use until the New Babylonian Empire in the sixth century BC, so we need to be careful when reading Old Babylonian numbers as the positional meaning of the symbols need to be taken in context. For example, without a zero we would have difficulty in distinguishing the numbers 18, 108 and 180. We are uncertain why the Babylonians chose to work with such a system; however, it is very efficient for calculations and has stood the test of time, principally through the use of base 60 for minutes and seconds in measurements of time and angles.

The hard evidence for Babylonian mathematics is in the form of clay tablets bearing cuneiform inscriptions. Their use was widespread, and hundreds of thousands of specimens have survived, from tiny fragments to whole tablets the size of a briefcase. Clay was plentiful, and so long as it remained moist one could wipe away a calculation and start another afresh. Once the clay had hardened, the tablet was either thrown away or used as building material. Arithmetical calculations were handled much as we would today. The Babylonians were prolific makers of mathematical tables, and they have left us some sophisticated tables of reciprocals, squares, cubes and higher powers – such higher powers being useful in calculating interest on loans. The use of mathematical tables is now largely a thing of the past because of the widespread use of calculators, but their importance in facilitating calculations has a long heritage going back to these clay tablets. The Babylonians were highly proficient in algebra, although questions and methods of solution were stated rhetorically – in words rather than symbols. They solved quadratic equations by what is essentially our method of 'completing the square', and their justification for the procedure was based on the fact that a rectangular area could be rearranged to form a square. Some higher-order equations were solved either by numerical methods or by simplifying them to known types.

In geometry they had procedures for finding the areas of plane figures, and many problems were solved algebraically. Irrational numbers, which give rise to an infinite decimal expansion, were handled numerically by truncating the fractional sexagesimal expansion. For example, in decimal notation, $\sqrt{5} = 2.236067\ldots$, the three dots signifying that the decimal expansion continues indefinitely. Truncating this to two decimal places yields the value 2.23, in contrast to the value 2.24, which is the closer approximation. Sometimes the truncated and the nearest approximation give the same result, for example, to three decimal places $\sqrt{5} = 2.236$ in both cases. There is no record of any discussion of the possible infinite nature of such expansions, but one tablet shows a very good approximation to $\sqrt{2}$, which in sexagesimal is given as 1: 24, 51, 10, and which is accurate to five decimal places. No derivation of this result is given, but a method named after Hero, a Greek mathematician of the first century AD, nearly two thousand years later, does yield exactly the same result. The Babylonians also made extensive use of the Pythagorean theorem a thousand years before Pythagoras was born.

Old Babylonian mathematics was both sophisticated and appropriate for the practical purposes of accountancy, finance, and weights and measures. Some of the problems that were tackled indicate that there was also a theoretical tradition, and we shall see the fruits of this when looking at Babylonian astronomy.

For a civilization that spanned some four thousand years, the Egyptians left precious little evidence of their mathematics. Papyrus is a fragile material and it is a miracle that any ancient papyri have survived at all. The two major sources are those known as the Rhind Papyrus and the Moscow Papyrus. There are also a handful of minor documents, and a number of illustrations on tombs and temples showing commercial and administrative problems that called for mathematical skills. The Rhind Papyrus was written in about 1650 BC by a scribe called Ahmes, who explains that he is copying an original which is two centuries older. The opening sentence claims that the text is 'a thorough study of all things, insight into all that exists, knowledge of all obscure secrets'. To us this may appear rather an exaggeration, but it may illustrate that the scribal art was the preserve of an inducted elite. The papyrus contains 87 problems and their solutions, and is written in everyday hieratic script rather than the elaborate hieroglyphic symbols preferred

➤ The Rhind Papyrus was discovered in the mid-nineteenth century, allegedly at Thebes. However, it was purchased in Luxor by A.H. Rhind and later sold to the British Museum, London, by his executors. This particular problem deals with finding the areas within a triangular shaped piece of land.

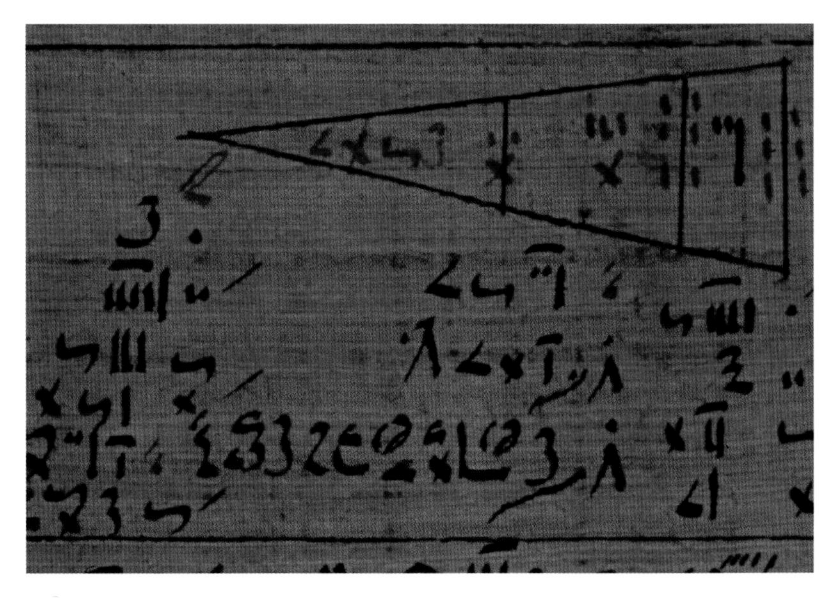

for ornamental writing. Most of the problems are computations, such as the division of a number of loaves of bread among a given number of men. There is also a method for finding the area of a right-angled triangle. All solutions are illustrated by worked examples; no general formulas are given explicitly. The Moscow Papyrus covers much the same ground, but includes also calculations of the volume of a truncated pyramid, or frustum, and of what appears to be the surface area of a hemisphere.

Two things immediately stand out as characteristic of the Egyptian use of numbers. The first is that all calculations are based on nothing more than addition and the two times table, and the second is their preference for unit fractions ($\frac{1}{2}$, $\frac{1}{3}$, etc.). Multiplication thus consisted in repeated doublings (and, if necessary, halvings), and then adding the appropriate intermediate results. For example, to multiply 19 by 5 the scribe would write

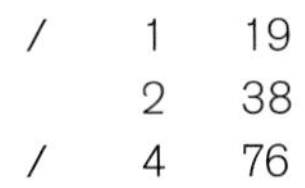

/	1	19
	2	38
/	4	76

▲ Relief on an Egyptian stele showing numerals, from the Temple of Karnak.

Then, since $1 + 4 = 5$, adding 19 and 76 gives 95, which is 19×5. Division would proceed similarly, but now there is the possibility of a fractional solution. This is where unit fractions come in. The Egyptian way of denoting a unit fraction was to place a bar over the number: thus $\frac{1}{5}$ was written as $\bar{5}$. There was no symbol for our $\frac{2}{5}$, or for any other fraction except $\frac{2}{3}$. The Rhind Papyrus starts with a table of fractions of the form $2/n$, where n is an odd number, decomposed into unit fractions. Thus $\frac{2}{5}$ is equal to $\frac{1}{3}$ and $\frac{1}{15}$, and wherever a problem had a solution which we would write as $\frac{2}{5}$, the Egyptian scribes would record it as $\bar{3}\overline{15}$. It is still difficult to see how this scheme would function on a practical level, though it obviously worked, and we await further discoveries to clarify its origins.

One possibility is that in numerical calculations to do with inheritances or the distribution of goods, unit fractions were used because they produced absolute accuracy rather than close approximation. The Egyptians had no currency, so transactions were made using other goods as a standard, the most common being bread and beer. This can be seen in the problem in the Rhind Papyrus of how to divide nine loaves among ten people. Now we would calculate that each would receive $\frac{9}{10}$ of a loaf, and distribute the loaves by cutting off a tenth from each loaf so that nine people would each receive one $\frac{9}{10}$ of a loaf, and one would receive nine $\frac{1}{10}$'s. The solution given in the papyrus is that $\frac{9}{10} = \frac{2}{3} + \frac{1}{5} + \frac{1}{30}$. This requires more cuts to be made, but in compensation each man received not only the same proportion of bread but also identical sized pieces.

Measures of volume had their own system of notation, consisting of parts of the hieroglyph representing the eye of Horus. We see here the dual role of a priestly caste as both administrators and religious functionaries. Horus was the falcon-god, and the eye was partly human and partly falcon. Each element of the hieroglyph represented a fraction from ½ to 1⁄64, and in combinations they could represent any number of 64ths. But the eye of Horus also had mystical significance. Horus was the only son of Isis and Osiris, and swore to avenge his father's death at the hands of his brother Seth. During one of their interminable battles Seth ripped out Horus' eye, tore it into six pieces and scattered them across Egypt. Horus returned the compliment by castrating Seth. Legend has it that the gods intervened and designated Horus as King of Egypt and tutelary god of the Pharaohs, instructing Thoth, the god of learning and magic, to reassemble Horus' eye. It thus became the symbol of wholeness, clear vision, abundance and fertility. The scribes, whose tutelary god was Thoth, used the talisman to symbolize the fractions of measures. It was told that one day an apprentice scribe observed to his master that the total fractions of the eye of Horus did not add up to unity, but rather to 63⁄64. The master replied that Thoth would make up the missing 1⁄64 to any scribe who sought and accepted his protection.

The king moreover (so they say) divided the country among all the Egyptians by giving each an equal square parcel of land, and made this his source of revenue, appointing the payment of a yearly tax. And any man who was robbed by the river of a part of his land would come to Sesostris [Pharaoh Ramses II, c. 1300 BC] and declare what had befallen him; then the king would send men to look into it and measure the space by which the land was diminished, so that thereafter it should pay in proportion to the tax originally imposed. From this, to my thinking, the Greeks learned the art of geometry; the sun-clock and the sundial, and the twelve divisions of the day, came to Hellas not from Egypt but from Babylonia.

Herodotus, *History*, II, mid-fifth century BC

Our knowledge of Egyptian mathematics is necessarily limited by a genuine lack of artefacts. It is therefore tempting to see the mathematics of the Egyptians as a step backward from the level reached by the Babylonians. But this is probably unwarranted, especially given their precision in pyramid-building and their management of such a vast empire. Some of the evidence we have tantalizingly presents us with what appear to be important results, such as the volume of a frustum, but it remains unclear whether this was an isolated result inspired by their interest in pyramids or part of a more advanced but sadly missing corpus of knowledge. The ancient Greeks widely acknowledged that their mathematics, especially their geometry, originated in Egypt. What strikes us most now is not the similarity between Egyptian and Greek mathematics but their enormous differences in style and depth and, we therefore assume, in understanding. It appears that the 'obscure secrets' of Ahmes remain so to this day.

chapter two

watchers of the skies

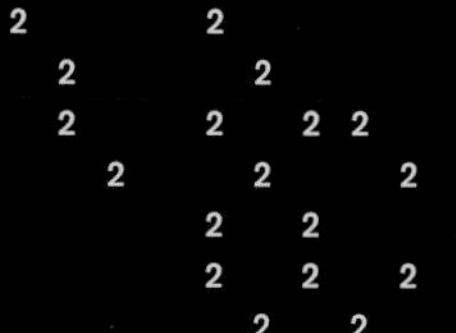

◄ The Aztec Sun Stone Calender, discovered in the sixteenth century, depicts Tonaituh, the Fifth Sun and symbol of our current epoch. Some believe that Aztec astronomical knowledge was inherited from earlier Central American civilizations such as the Olmecs and Mayans.

Much early mathematics was developed for trade and agriculture, but there were also links with religious observances and the motions of the heavens. The construction of calendars was essentially the work of astronomer-priests, and mapping the skies required special mathematics to be developed. As most ancient cosmologies were geocentric the term 'planet' refers to the Sun, Moon and the five visible planets, with Uranus, Neptune and Pluto discovered relatively recently. Across the Earth, various civilizations recorded the motions of the heavenly bodies and constructed calendars, and all had to find a way to reconcile the two most important temporal cycles – the lunar month and the solar year.

The Mayan civilization of Central America, which can be traced back to 1000 BC, had its classical period from AD 300 to 900. Few documents survived the Spanish conquests from 1519 onwards (the most important is the manuscript known as the Dresden Codex, which contains astronomical tables), but thankfully the Maya also left us carvings. Every twenty years they erected stone stelae, or pillars, recording the date of construction, principal events of the previous twenty years, and the names of nobles and priests. The hieroglyphs used for these and other inscriptions were stylizations of Mayan divinities. But for numbers they often used a notation now known as 'dot and dash'. In this concise place-value system of numeration, a dot represented 'one' and a horizontal dash represented 'five', with a symbol for zero that looked like a shell. The system seems to have been used from about 400 BC, and was essentially vigesimal, or base 20, apart from an anomaly in the third place. A true vigesimal system would have place values in the sequence 1, 20, 20^2, 20^3, and so on, but the Mayan system uses the sequence 1, 20, 18×20, 18×20^2, and so on. This complicated certain calculations, but from the fact that $18 \times 20 = 360$ we can see the importance the Maya attached to their calendar.

The Maya had three calendars. The sacred year consisted of 260 days in two overlapping cycles: one of the numbers 1 to 13, and the other a 20-day cycle of divinities. Thus every day in the sacred year was uniquely defined by a number and deity. This calendar was of little use to farmers, and a civil year of 365 days was also used, composed of 18 months of 20 days plus an extra 5 days known as the 'period without a name'. The hieroglyph for this last period represented chaos and disorder, and anybody born on one of these days was supposed to be cursed for life. A third calendar used for 'long counts' was based on a chronology dating back to 12 August 3013 BC and cycles of 360 days. There were further sacrificial cycles of 4, 9 and 819 days, so much of the scribes' time was spent calculating calendars and significant dates. Without any evident use of fractions or trigonometry, the Maya were able to make very accurate predictions of cycles based on an enormous wealth of accumulated astronomical observations. For example, Mayan astronomers claimed that 149 lunar months equal 4400 days, which is equivalent to a lunar month of 29.5302 days – very close to our currently accepted value of 29.530 59 days. The Dresden Codex includes tables of lunar and solar eclipses, and of predicted positions of Venus as the 'morning star' and the 'evening star'. Little else is known about Mayan mathematical astronomy.

The Egyptian calendar employed exactly the same scheme as the Mayan, with 12 months of 30 days and 5 additional days at the end of the year. It was the Egyptians who first divided the day into 24 units, though it is not clear when the hour became of fixed duration. They used what can be termed 'seasonal' hours, dividing daytime and night-time into 12 units each, which varied as the lengths of daylight and darkness changed during the course of the year. The Egyptians had their own set of small constellations, the 'decans', whose risings were ten days apart. In Hellenistic times these were combined with the Babylonian zodiac so that each zodiacal constellation, spanning 30 degrees of the sky, was further divided into three decans. The decans are depicted on the ceilings of royal temples and on coffin lids from the Middle Kingdom (*c.* 2100–1800 BC). But it has proved difficult to identify the decans with the known stars. The only exception is the star Sirius, whose rising at a certain time of year heralded the annual flooding of the Nile, which was essential to irrigation. On later tombs we find a more elaborate portrayal of stars on a grid system, and there is a Demotic papyrus from the Hellenic era which has helped in deciphering the inscriptions. However, it seems that the craftsmen who added these depictions to the tombs took great artistic liberties in interpreting the astronomical information, for the initial sketches that underlay the final images were actually more accurate. We have no written documentation about astronomical observations or the production of tables by the Egyptians; even Ptolemy, who quotes his sources of ancient astronomy, does not cite any Egyptian data.

▲ A modern reconstruction of a Byzantine portable sundial and calender of the sixth century. The back of the instrument reveals a sophisticated mechanism of gears, based upon those discovered with the original.

From the fall of the Assyrian Empire to Hellenistic times, the Babylonians developed an effective predictive astronomy. Ptolemy mentions that from the eighth century BC there were complete lists of lunar eclipses, but that there was little reliable planetary data. The Babylonian calendar was purely lunar, the first day of the month falling on the first sighting of the crescent Moon, each day lasting from one sunset to the next. They were therefore most interested in predicting the appearance of the crescent Moon, and whether a month was to be 29 or 30 days long, depending on the relative positions of Sun and Moon. Similarly for the planets, much of the interest was in first sightings; especially important in the early tablets is the planet Venus. For the purposes of producing ephemerides, as tables of planetary positions are called, the zodiacal

▲ Medieval astronomers using an astrolable, an invention ascribed to the ancient Greeks but developed to perfection by Arab scientists and mathematicians.

region was divided into three zones of twelve named constellations, and planetary positions were given with reference to the stars. There are also tables of the rising and setting times for the constellations. Ephemerides are found from the Seleucid period, especially for the Moon but also for other planets.

One of the greatest achievements of this period was the analysis of the apparent motions of the Sun and Moon, essential in determining the beginning of each month. The Babylonians established that the angle between the horizon and the ecliptic, the apparent path of the Sun across the sky, varies throughout the year. Also, the Moon's path deviates periodically from the ecliptic by about 5° either side. On top of this, both bodies move at a variable rate. These periodic changes, which are sinusoidal variations, were approximated to a high degree of accuracy by so-called 'zigzag functions'. These were handled arithmetically as rising and falling number sequences. Many Babylonian tablets showing exercises in arithmetic progressions could be preparations for creating solar and lunar tables. These tables could be used to predict the appearance of the crescent Moon, which depends on the relative positions of the Moon and the Sun, up to three years in advance. From the evidence we have, it seems that methods of arithmetical interpolation were used to smooth out the planetary paths from the key observations. The Ptolemaic theory (see below) took the opposite approach, seeking to build the most accurate planetary model possible from which to derive planetary positions.

It is not clear what the later Babylonian planetary theory was; early records indicate a geocentric universe with circular planetary orbits. In the Hellenic world, Aristarchus (*c.* 320–*c.* 250 BC) suggested a heliocentric system, probably based on his calculations that the Sun was by far the largest heavenly body. But the theory found little favour at the time, and was not to resurface until the sixteenth century. Greek planetary theory was dominated by the view of Aristotle (384–322 BC), that the planets have perfect motion, following circular orbits of constant speed. This philosophical position was maintained in the face of clear observational evidence of variable speed, retrogressions and variations in the apparent brightness of the planets. Such discrepancies between theory and observations were resolved by introducing epicycles: a planet no longer orbited the Earth itself, but around an epicycle, a circular orbit whose centre moved along a deferent, a circle centred on the Earth. By this artifice, the constant speed of the planet was transformed into its apparent variable speed, at the same time keeping the planets within circular 'shells' if not perfectly circular orbits. This system found its most complete expression in the work of Ptolemy.

Before looking at Ptolemy, we must mention the most famous of his predecessors, Hipparchus (190–120 BC), a mathematician from Nicea, in modern Turkey. He was considered the greatest astronomer of his day and is credited with founding astronomy on Greek geometric principles. He employed the division of the circle into 360 degrees as the basis of trigonometry, with each degree further divided into sixty minutes. His

➤ A diagram of the Ptolemaic cosmos from the *Breve compendio de la esfera y de la arte de navigar*, by Martin Cortes de Albacar (1551), a treatise on cosmology and navigation.

Primera. Fol.xiij.

¶ Capitulo sexto de la inmutabilidad de la tierra.

Opinion d' ātiguos a cerca de mouer se la tierra.

Los pithagoricos y otros naturales antiguos como trae Aristotiles sintieron q̃ la tierra se mouia no del mouimiēto recto mas cerca del medio y circularmente el qual error assi el philosopho como los astronomos por euidētes causas y demōstraciōes manifiestas cōfūdē y reprueuā. Por q̃l mouimiēto circular es ppio

Aristo. 4 phisi.

lucretius philo.

A v

treatise on the subject included a table of chords (a chord is essentially double the sine of half an angle) whose values were calculated for a circle with a radius of 3438 minutes, this being the radius necessary to ensure a circumference of 360×60 = 21600 minutes. These tables, which are very similar to tables found in Indian mathematics, enabled Hipparchus to describe the positions of heavenly bodies more accurately. He modelled the motion of the Sun and Moon using a geocentric system of epicycles. Hipparchus admitted that his data was not accurate enough for him to speculate on the orbits of the other planets. Unfortunately, only one minor work of his survives and he, like many other Greek astronomers, was to be overshadowed by Ptolemy.

Claudius Ptolemy (*c.* AD 85–*c.* 165) lived in Alexandria, and we know he started making astronomical observations there on 26 March 127. Very little is known of his family background or the exact dates of his birth and death. He left several writings, the most famous of which is entitled *Syntaxis* ('The Mathematical Collection'). The high regard with which this work was held when translated into Arabic, in about AD 820, earned it the name *al-Majisti* ('The Greatest'), and it subsequently became known as the *Almagest* when translated into Latin. Ptolemy's *Almagest* did for astronomy what Euclid's *Elements* did for geometry, thereby consigning to obscurity much of the work that went before, save for Ptolemy's own historical commentary. It begins with some preliminaries on trigonometry and chords, followed by a detailed theory of the Sun's motion, assigning it a circular orbit but placing the Earth slightly away from the centre of the orbit, which occupied a position he called the eccentric. In his theory of the Moon's motion, Ptolemy borrows extensively from Hipparchus and improves on his epicycle model. Combining the motions of the Sun and Moon, Ptolemy then discusses lunar and solar eclipses. There follows a demonstration that the sphere of the fixed stars, the outer shell of the Hellenistic cosmos, is indeed stationary, since Ptolemy's own observations of the stars agreed with those of Hipparchus, made some two hundred years earlier. After his extensive catalogue of over a thousand stars, Ptolemy comes to the orbits of the remaining five planets. An ingenious construction involved a point known as the equant, which was the same distance from the Earth as the eccentric, but on the opposite side. Ptolemy constructs a planet's cycle so that it has a constant speed about the equant. With cosmology having wandered so far from Aristotelian perfection, we may wonder why the philosophical restrictions were not wholly abandoned. But an Earth orbiting the Sun was incompatible with the contemporary understanding of terrestrial dynamics: it was still believed that we would fly off the surface of a moving Earth. The Ptolemaic model is by far the most successful attempt at a predictive astronomy ever developed, reproducing the apparent motions of the planets, including retrograde loops. Any discrepancies with observations were usually within the margin of error of the observational techniques. The system was not called into serious question until the sixteenth century, by which time Ptolemy's *Almagest* had been the established authority for 1400 years.

chapter three

the pythagorean theorem

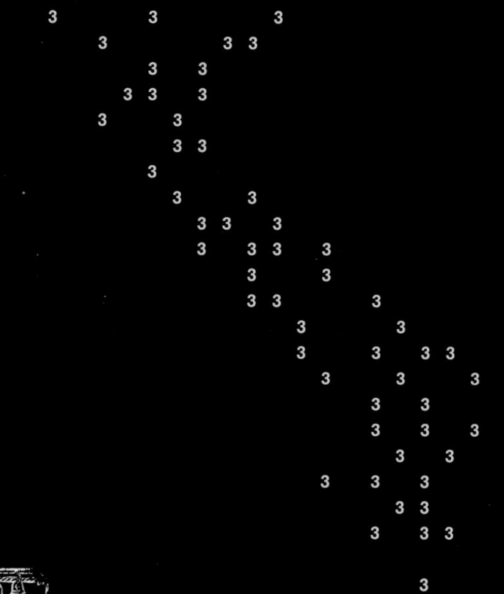

◄ Medieval woodcut celebrating the Pythagorean contribution to music. The discovery of the relationship between numbers and musical intervals has continued to resonate in the concept of the harmony of the spheres.

▼ The Babylonian tablet now known as Plimpton 322 has been one of the most analyzed mathematical artefacts from antiquity. It is now considered to be a table of fractional Pythagorean triples, over a thousand years before Pythagoras was born.

Virtually everyone has come across one mathematical theorem at school. It now bears the name of Pythagoras, but it was widely known in antiquity well before Pythagoras was even born. The existence of the theorem gives us the opportunity to compare the mathematical styles and preoccupations of some of the ancient mathematicians in different cultures.

The pythagorean theorem: for a right-angled triangle, the sum of the squares on the two shorter sides equals the square on the longer side. It is possible to form such triangles with integral (whole-number) sides, the most famous being the 3,4,5 triangle. There are an infinite number of such pythagorean triples, as they are called, for example, the 5,12,13 and 7,24,25 triangles, which were also known in antiquity.

One of the most fascinating Babylonian mathematical artefacts is the tablet now known as Plimpton 322, kept at the University of Columbia, New York. It bears four columns and fifteen rows of numbers, and appears to be incomplete; it may well be part of a larger tablet which became broken. It is now generally accepted that this tablet presents derivations of fractional Pythagorean triples. Such a sophisticated procedure must mean that the Babylonians understood the Pythagorean theorem as early as 1800–1650 BC, more than a thousand years before Pythagoras. This interpretation has been borne out

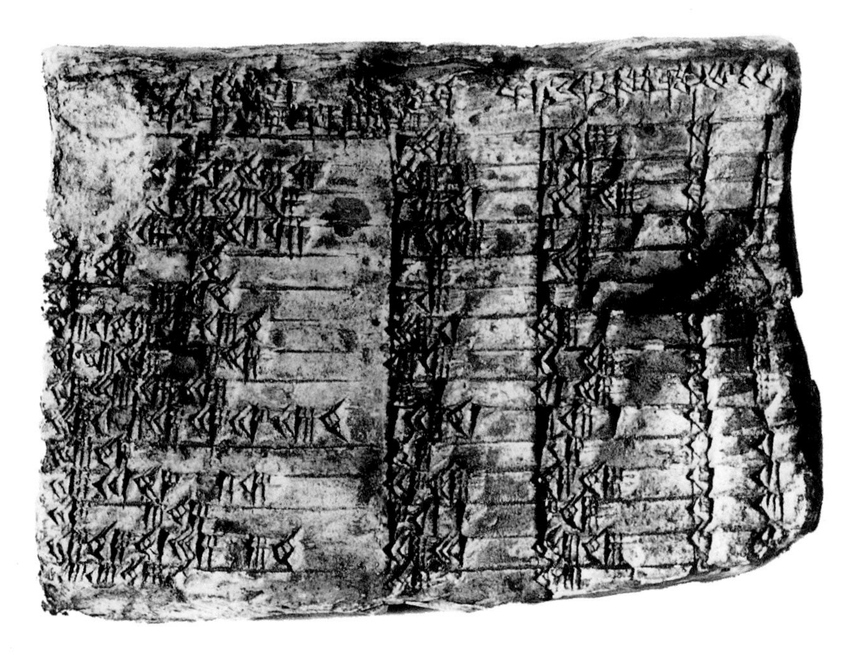

by a further tablet found near Babylon dating from the same era, now one of the oldest examples of the theorem. The Babylonians used the rule for geometric calculations and to find solutions to algebraic equations, though such algebra was rhetorical rather than symbolic. Some have speculated that these Babylonians may well have made a start in developing an early form of trigonometry.

The Vedic period of Indian civilization is generally reckoned as starting around the beginning of the first millennium BC. This period saw the establishment of Hindu culture and religion through scriptural texts such as the *Vedas* and *Upanishads*, and rules of social conduct such as the *Code of Manu*. The mathematics of the period is recorded in the *Sulbasutras*, part of the appendices to the *Vedas*, and not surprisingly a large part is devoted to the mathematics needed to ensure conformity with the rules of ritual. The term *sulba* came to mean the cord or rope used to measure the dimensions of altars. Three versions of these texts have come down to us, the earliest probably recorded between 800 and 600 BC. A simplified Pythagorean theorem is stated by Baudhayana as: 'The rope which is stretched across the diagonal of a square produces an area double the size of the original square.' A later, more general proposition is given by Katyayana: 'The rope of the diagonal of a rectangle makes an area which the vertical and horizontal sides make together.' No proof is given, but a number of practical applications are described. There were rules that demanded altars to be constructed to have areas which were multiples of those of other altars of the same design, and the imperative need for accuracy meant that geometrical methods were preferred to numerical ones. For example, if one needs to double the area of a square, it is quite simple to construct a square whose sides are the diagonal of a given square, rather than having to calculate that the new side will be larger by a factor of $\sqrt{2}$. Indians had excellent methods for estimating $\sqrt{2}$, but for religious reasons absolute accuracy was demanded – an estimate was not good enough.

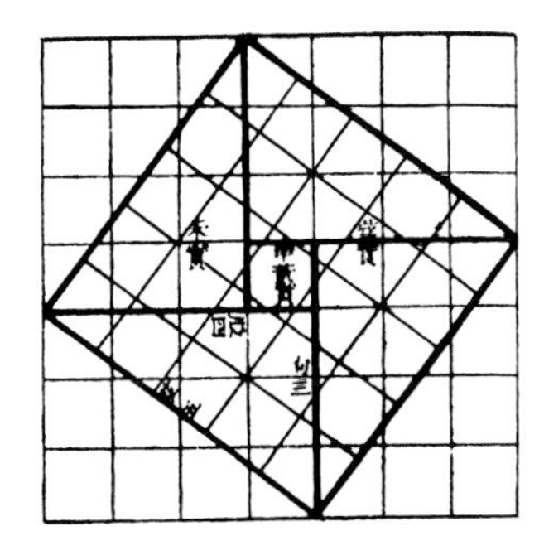

▲ Proof of the Pythagorean Theorem as illustrated in the Chinese *Zhoubi suanjing* (the original is thought to date back to *c.* 400–200 BC). This proof relies on the 3,4,5 triangle, a Pythagorean triple well known in antiquity, where $3^2 + 4^2 = 5^2$.

The earliest recorded Chinese mathematical text is the *Zhoubi suanjing*, ('Zhou Dynasty Canon of Gnomonic Computations'), thought to have been written between 500 and 200 BC but based on a text from the Shang Dynasty, possibly five hundred years earlier. As the name of the text suggests, it deals mainly with astronomy; it also includes some preliminary instructions on arithmetic and geometry. It was written during the time known as the Warring States, a period of instability between the Zhou and Han Dynasties, perhaps by one of the many itinerant philosophers whose precious advice was sought by feudal lords. The most famous such philosopher was Confucius, whose philosophy of unity and stability can be seen as a reaction to the turbulent times.

The first section of the 'Gnomonic Computations' is a dialogue between Duke Zhou Kung and a notable called Shang Kao in which they discuss properties of right-angled triangles. The Pythagorean theorem, known as *gougu*, is stated followed by a geometric

➤ The Pythagorean theorem as discussed in an Arab text. The proof given is after Euclid with the characteristic 'windmill' diagram illustrating the proof geometrically.

demonstration. A process called 'piling up the rectangles' is employed, with a diagram showing the method for the triangle with sides 3, 4 and 5, the smallest Pythagorean triple. The extension to other lengths must have been obvious to the reader, but an explicit general statement was left to commentators of the third century AD. One such writer, Liu Hui, gives a second geometric proof using the principle of 'out–in complementarity' in which the two smaller squares are cut in such a way as to construct the larger square. The rule that $gou^2 + gu^2 = xian^2$ (our $a^2 + b^2 = c^2$) was then used in numerous problems. The theorem was of great importance in Chinese mathematics as it formed the basis of other methods such as the extraction of square roots and solutions of quadratic equations. One classic problem, known as the 'broken bamboo', was later to reappear in European books, possibly charting a westward migration of Chinese mathematics through India and the Arab world.

We finally come to the legend that is Pythagoras (*c.* 580–*c.* 500 BC). It is perhaps fitting that Pythagoras was a near-contemporary of Buddha, Confucius, Mahavira, Laozi, and probably Zoroaster. His blend of mathematics and mysticism has resonances even in the present day, most often in its third-century BC development, Neoplatonism. The real Pythagoras remains unknown. References to him are often partisan, and even Aristotle, barely two hundred years later, was unable to furnish us with a clear picture of the man. The importance of Pythagoras and his followers is in their philosophy of mathematics. The belief in the primacy of mathematics as the one true source of knowledge has come down to us through philosophers and mathematicians such as Plato, Plotinus, Iamblichus and Proclus (AD 411–485), and is a cornerstone of the Neoplatonism that has found numerous expressions in Western thought.

After studying under the Egyptians and Chaldeans, Pythagoras settled in Crotona, in what is now southern Italy, and founded a school there. This school was more like a secret society or cult, some of their knowledge being passed on only to a select group of initiates. The Pythagoreans lived a communal life, with a strict code of morals and conduct. This included a belief in metempsychosis, or transmigration of the soul, and strict vegetarianism. As he left no writings, we can do no more than speculate as to which mathematical results can be ascribed to Pythagoras himself. There are often references to the Pythagoreans, suggesting that members of the school later relaxed their master's ban on publishing. One of the key teachings of the Pythagorean school was that numbers were everything, and that nothing could be conceived or known without numbers. Their most venerated number was ten, or the *tetractys*, being the sum of 1 + 2 + 3 + 4. These are the number of points needed to generate the dimensions of the universe: 1 is the dimensionless point, and generator of the other dimensions; 2 points can be joined to create a line which has one dimension; 3 points can be joined to make a two-dimensional triangle; and 4 points can be joined to make the three-dimensional tetrahedron. The *tetractys* became the symbol of the Pythagoreans, who went further

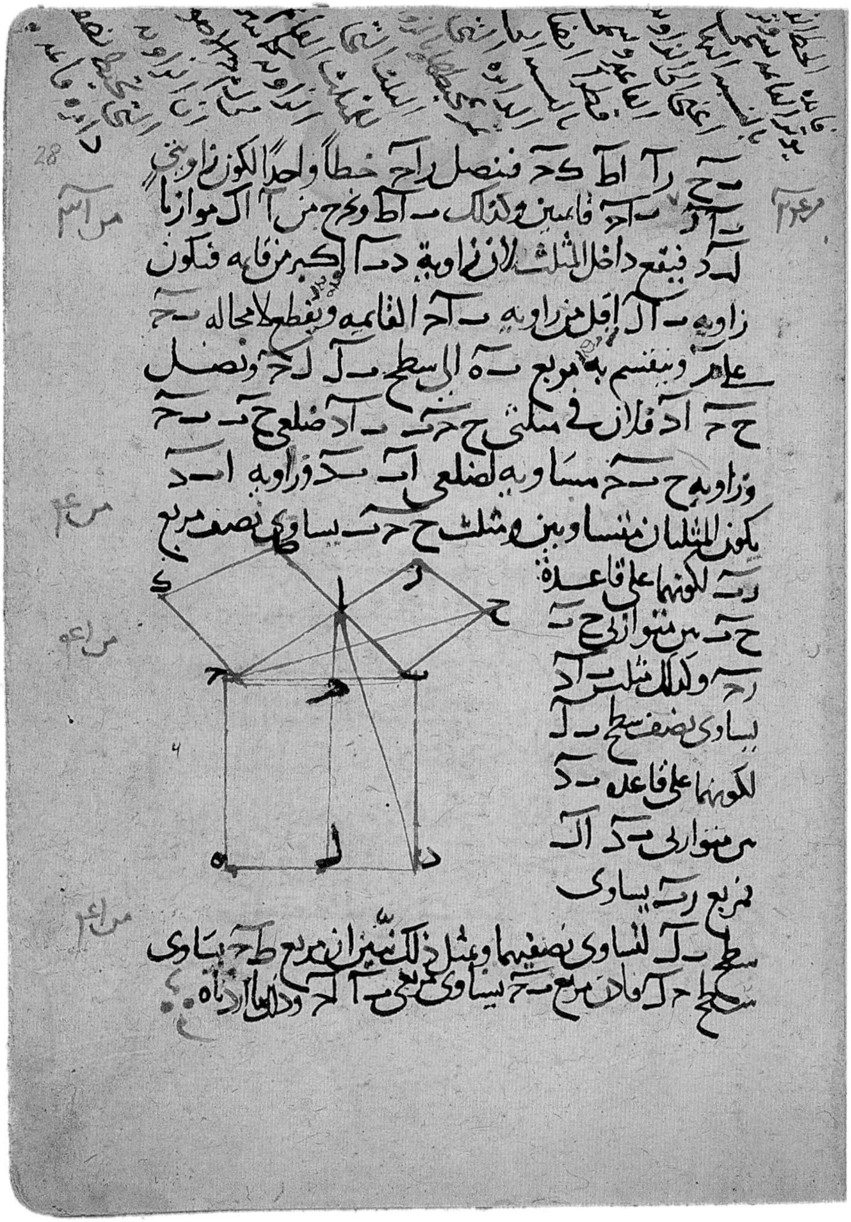

than any previous number-mysticism in constructing a universe in which numbers were assigned both philosophical and revelatory roles. The Pythagoreans are also credited with the numerical analysis of music, and here the *tetractys* symbolized key ratios between notes, starting from the ratio 1:2 for the octave. The whole concept of the harmony of the spheres comes from this numerology of music, which was to so influence Kepler's planetary model more than two thousand years later.

But the name of Pythagoras is best known for the theorem that now bears his name. As we have seen, the theorem was actually well known throughout antiquity. It is claimed that Pythagoras learnt of the rule from a civilization we have not mentioned in this connection – the Egyptians. In fact, Greek literature makes various references to Egypt as the source of their knowledge in geometry, and it is unfortunate that we have no Egyptian documents illustrating the Pythagorean theorem. Aristotle ascribes to the Pythagoreans the first proof that $\sqrt{2}$ is irrational. Given a right-angled triangle with base and height of length 1, then the hypotenuse will be $\sqrt{2}$ in length. In the language of Greek mathematics, the Pythagoreans sought to express the ratio of the hypotenuse to the unit length, or $\sqrt{2}:1$ as we would now write, as a ratio of whole numbers. In contrast to, for example, the (3, 4, 5) triangle, where any pair of sides will be in a ratio of whole numbers, this proved impossible to achieve with our given triangle. The hypotenuse and the unit side were said to be incommensurable, that is, given a ruler with any equally-spaced markings then the two sides cannot both be measured *exactly* by it. Given that the unit side is rational, then the hypotenuse is irrational with respect to it. The historian Diogenes tells that this discovery was made by Hippasus of Metapontum, a member of Pythagoras' school, and that his colleagues took him out to sea and cast him overboard for destroying their faith that everything could be expressed as whole numbers and their ratios. This story is now considered doubtful, but the relationship between commensurable and incommensurable lengths, and that between rational and irrational numbers has been of great importance in mathematics. Indeed, a *definition* of irrationals in terms of rationals would not be settled upon for over two thousand years (Chapter 19).

Most striking about the Greek handling of the Pythagorean theorem is their method of proof, which is found at the end of Book I of Euclid's *Elements*. It is a very general geometric proof employing a sequence of constructions which transform the two smaller squares into two rectangles, which fit together to form the larger square. It is presented without any reference to numerical values, and the characteristic 'windmill' diagram that goes with the proof is later found in the mathematics of many Eurasian cultures. Indeed, Proclus was to comment that, 'while I admire those who first observed the truth of this theorem, I marvel more at the writer of the *Elements*'. Nevertheless, it is Pythagoras' name that has been wedded to the theorem, and the attraction of the Pythagorean ideal of a mathematical universe lives on.

chapter four

the elements

202 *EUCLIDIS Elementorum*

a 3. ax. 10.
b 1. def. 10.
c 16. 10.

1. *Hyp.* Si fieri poteſt, ſit D ipſarum AC, AB communis menſura. [a] ergò D metitur AC — AB (BC). [b] ergò AB ⊏⊐ BC, contra Hypoth.

2. *Hyp.* Dic AB ⊏⊐ BC. [c] ergò AC ⊏⊐ AB, contra Hypoth.

Coroll.

Hinc etiam, ſi tota magnitudo ex duabus compoſita, incommenſurabilis ſit alteri ipſarum, eadem & reliquæ incommenſurabilis erit.

PROP. XVIII.

Si fuerint duæ rectæ lineæ inæquales AB, GK; *quartæ autem parti quadrati, quod ſit à minori* GK, *æquale parallelogrammum* ADB *ad majorem* AB *applicetur, deficiens figurâ quadratâ, & in partes* AD, DB *longitudine commenſurabiles ipſam dividat, major* AB *tanto plus poterit quàm minor* GK, *quantum eſt quadratum rectæ lineæ* FD *ſibi longitudine commenſurabilis: Quòd ſi major* AB *tanto plus poſſit, quàm minor* GK, *quantum eſt quadratum rectæ lineæ* FD *ſibi longitudine commenſurabilis; quartæ autem parti quadrati, quod ſit à minori* GK, *æquale parallelogrammum* ADB *ad majorem* AB *applicetur, deficiens figurâ quadratâ, in partes* AD, DB *longitudine commenſurabiles ipſam dividet.*

a 10. 1.
b 28. 6.
c 8. 2.
d conſtr. & 4. 2.

[a] Biſeca GK in H; & [b] fac rectang. ADB = GHq: abſcinde AF = DB. Eſtque ABq [c] = 4 ADB [d] (4 GHq, vel GKq) + FDq. Jam primò

◄ A page from Newton's copy of Euclid's *Elements* with marginal notes in his own hand.

▼ Frontispiece of Edmond Halley's edition of the works of Apollonius, published in 1710, illustrating the classical story of the philosopher Aristippus being shipwrecked off the coast of Rhodes. He was confident of the civilized nature of the locals after seeing geometric figures drawn in the sand.

The Greeks enter history as invaders from the north who settled in the land between the Ionian and Aegean Seas. They showed an insatiable desire to learn from their more ancient neighbours and, more importantly, to transcend the received wisdom of the Egyptians and Mesopotamians. The Greek or Hellenic world was bound by cultural rather than racial bonds. It can be split into two broad phases, with Alexander the Great marking the transition; for the purposes of mathematics these phases may be termed the Athenian and Alexandrian periods.

The first Olympic Games were held in 776 BC, by which time Greek literature could already boast the works of Homer and Hesiod, but of their mathematics we know nothing until the sixth century BC. The title of first Greek mathematician probably goes to Thales of Miletus (*c.* 624–548 BC), who is reputed to have given the first demonstrations of various geometric theorems, thus presaging the great deductive system of Euclid. But our knowledge of Greek mathematics and of this period in general is prone to a certain amount of historical noise. Not only do we have no copies of the writings of the time, but we are forced to rely on commentaries written as much as a thousand years after the events they purport to describe.

The fourth century BC saw Athens as the centre of the Mediterranean intellectual world, with the founding of Plato's Academy, and later his former student Aristotle's Lyceum. The role of Plato in the history of mathematics is still a topic of debate. He left no formal mathematical writings of his own, but has been highly influential in the philosophy of mathematics. In the *Republic* he urges that mathematics should be one of the core subjects for future rulers, and in the *Timaeus* we find a kind of reformed Pythagoreanism, with the Platonic solids associated with the four elements and the dodecahedron a symbol for the whole universe. The influence of Aristotle's philosophy has not been entirely beneficial for mathematics. His demand for logical exposition had a positive effect, but a refusal to countenance the use of infinities and infinitesimals, coupled with his belief that perfect motion took place in circles and straight lines because these were perfect figures, may be seen as less than beneficial.

The Academy and the Lyceum were both important centres of mathematical teaching and research. Aristotle was a tutor of Alexander the Great, whose empire at its height stretched all the way to northern India. On his death, Alexander's vast empire was fractured by his warring generals. But one part of the splintered empire emerged as a centre of learning under the enlightened rule of Ptolemy I – the new city of Alexandria, with its Museum and its precious Library. Alexandria would largely eclipse Athens in the second period of classical Greek civilization, and what has become known as the Golden Age of Greek mathematics.

The most important single work in Greek mathematics is undoubtedly the *Elements*, by Euclid (*c.* 325–265 BC). Despite such fame, very little is known about Euclid's life, not even his birthplace. We do know from a passage by the later commentator Proclus

➤ A medieval Latin copy from the Arabic, usually ascribed to Adelard of Bath, but possibly a further copy. The propositions here are merely stated with the aid of diagrams. The only commentaries on the proofs are found in Book 1 of this manuscript, perhaps supporting the view that medieval learning of geometry was largely confined to the simplest books of the *Elements*.

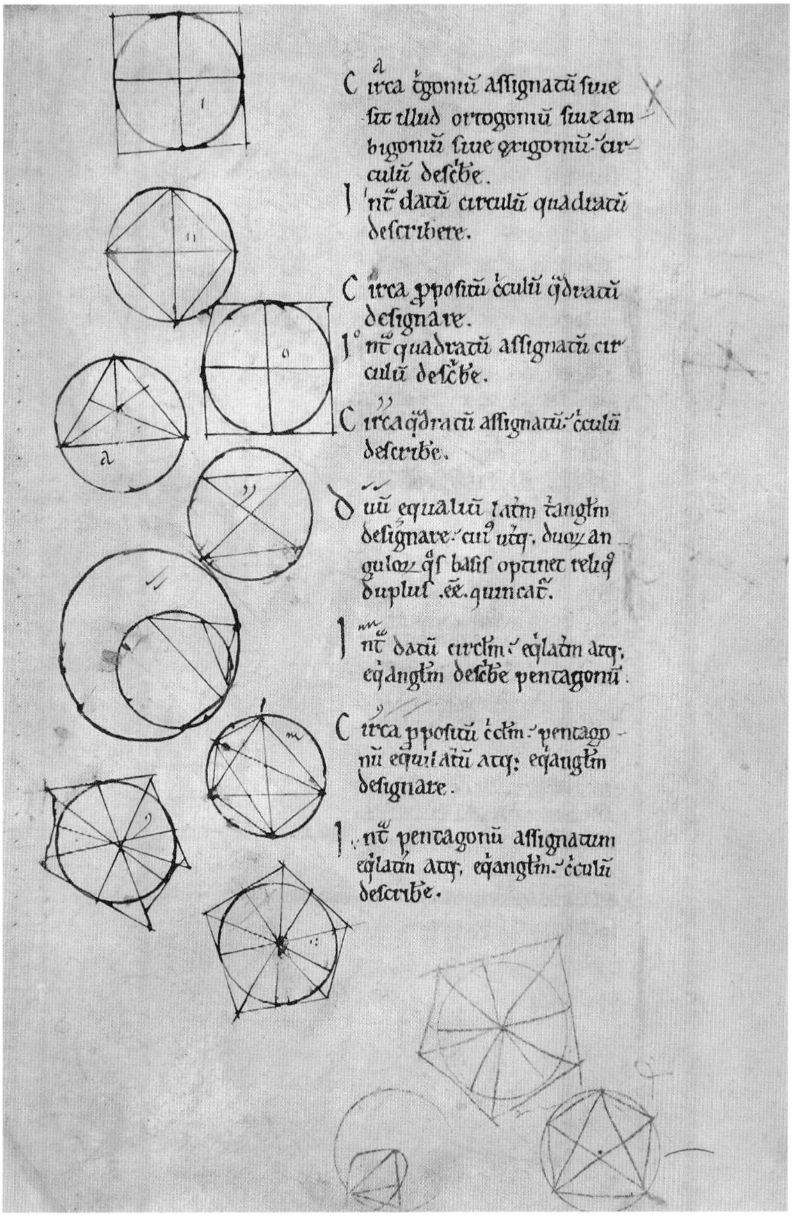

(*c.* 410–485) that Euclid taught in Alexandria during Ptolemy's reign, and that it was in answer to a request from the emperor for a short cut to learning geometry that he replied, 'There is no royal road to geometry.' The fame of the *Elements* has sometimes eclipsed the fact that Euclid wrote numerous other works on optics, astronomy, mechanics and music. But the *Elements* became the standard textbook on geometry for centuries to come, making previous such books so redundant that no copies of them have survived. As with all textbooks, much of the *Elements* is not original work, but it is Euclid we must thank for bringing together results from other sources, and presenting them within a structure which has become the accepted model for a logical, deductive system of theorems and proofs. The *Elements* is not a compendium of all Greek mathematics, only the elementary part of the subject. Calculation skills are not included, neither are advanced mathematical subjects such as conic sections.

The *Elements* is divided into 13 books, and covers the subjects of elementary plane geometry, the theory of numbers, the theory of incommensurables and solid geometry. It begins abruptly with a list of 23 definitions, such as 'A point is that which has no part' and 'A line is breadthless length.' This is followed by five postulates and five 'common notions', the notorious fifth postulate having a history all of its own. In fact, each section of the book opens with further definitions pertinent to the new topics being considered. To Euclid, the definitions were more self-evident than the postulates, though to us they would all be considered on a par with axioms. The postulates tend to be procedural, such as 'To draw a straight line from any point to any point', whereas the fourth definition states that 'A straight line is a line which lies evenly with the points on itself.' Taken as a whole, we see here the restriction of geometry to ruler-and-compass methods of construction. These two simple tools served as the logical generators of the whole system, the circle and straight line being the most perfect figures. The Greeks did use other so-called 'mechanical' methods of construction, but the *Elements* does not deal with these.

Books I to IV deal with geometric constructions of plane figures, including quadrilaterals, triangles, circles, and polygons constructed with the aid of circles. It has been argued that parts of these books, especially Book II, hint at a kind of algebraic geometry, where geometric constructions serve the same function as algebraic manipulations. Whether this is true or not, it does appear that at least in these early theorems Euclid is concerned with wholly geometric concepts. The term 'magnitude' is used throughout to denote any geometrical object – a line segment or figure – and the theorems are about the constructions and the relationships between these magnitudes. There is no appeal to numerical concepts such as length, so that, for example, a square is treated as a geometric construction arising out of a line segment. Nowhere does Euclid state that the area of such a square is the product of its sides – this comes much later. Magnitudes are thus the most elementary concept in the *Elements*, the foundation upon which the rest of the work is built. In this context it is interesting that the proof of the Pythagorean theorem proceeds by reconstructing figures, when an appeal to the actual areas involved may have given a very different proof.

Book V is on the general theory of proportionals, as first expounded by Eudoxus. A member of Plato's Academy, Eudoxus of Cnidus (*c.* 408–355 BC) was one of the most famous mathematicians of his time. He is credited with two fundamental discoveries: the theory of proportionals and the method of exhaustion. The apparent crisis of incommensurables was largely saved by retaining an ability to manipulate their products and ratios through Eudoxan proportions. Euclid, in fact, cites a number of different rules for proportions and their conditions of use. The preference for ratios over fractions had some advantages. One could state a rule such as 'the ratio of the areas of circles are in proportion to the squares on their diameters' and employ this in a variety of theorems without recourse to using π, which is irrational. Also, a ratio of magnitudes of the same type is dimensionless, and so can be put in proportion with other ratios, as in the example above. Thus the ratio was the most basic relationship between magnitudes, and the theory of proportionals enabled different ratios to be compared with one another. Book VI covers rules on similar figures and contains a generalization of the Pythagorean theorem not restricted to squares on the sides of the triangle, but expanded to include any constructible figure. Thus if we construct semicircles with each side of the triangle as a diameter, then the sum of the two smaller semicircles is equal to the larger one.

The theory of numbers is treated in Books VII to IX. To Euclid, 'numbers' meant whole numbers. From the definitions in Book VII, we see that the context of handling numbers is essentially geometric. Euclid says that 'The greater number is a multiple of the lesser when it is measured by it' and that the product of two numbers is the area of a rectangle. There is also the celebrated rule known as the Euclidean algorithm for finding the highest common factor of two numbers or, in the words of Euclid, 'the greatest common measure between two magnitudes'. In Book IX we find the famous proof which, in modern garb, states that there are an infinite number of primes. In fact Euclid expressly avoids appeals to the infinite. He states that 'Prime numbers are more than any assigned multitude of prime numbers' and proceeds to prove this for only three given primes, with the required extension to any assigned multitude left implicit. This book also presents a rule for constructing perfect numbers. A perfect number is one for which the sum of its factors is equal to the number itself. The first perfect number is 6, and the second is 28 (its factors are 1, 2, 4, 7, and 14, which sum to 28).

Book X is a detailed analysis of various irrational lengths, and it is here that we find the idea of incommensurability between general magnitudes honed down to the concept of irrationality between lengths (and squares). Given an assigned straight line, defined as being rational, then any straight line incommensurable with it is said to be irrational. Lengthy proofs are given for all the different types of irrationality, from simple square roots to multiple roots such as $\sqrt{(\sqrt{a} + \sqrt{b})}$. A discussion about ways of expressing

▲ The Archimedes palimpsest is a tenth-century Byzantine manuscript which has at some stage been partially cleaned and over-written with a liturgical text, common practice when paper was in short supply. The faded text has been electronically enhanced and includes a previously unavailable Archimedean work, *The Method*.

irrationals numerically sheds some interesting light on the problems encountered. A notation did exist based on the Euclidean algorithm, but although it yielded a useful expression for an individual irrational there was no simple procedure for expressing the sums or products in the same notation. One curiosity is Lemma 1 (a lemma being a preliminary theorem), which finds two square numbers whose sum is a square – the Pythagorean theorem presented as number theory with no reference to the proof given at the end of Book I. It is in this Book that there is the strongest suggestion that such numeric-geometric procedures are a preliminary to more advanced problems, such as the finding of areas, the quadrature problems. It may also be noted that the irrationals dealt with can all be constructed with ruler and compass – there are no cube roots, for example. The lengthy classification of irrationals becomes meaningful in the last sections of the *Elements*, where they reappear in connection with the regular solids.

The final three Books of the *Elements* deal with solid geometry and use the Eudoxan method of exhaustion as a rigorous way of finding areas and volumes through repeated approximations. Archimedes ascribed to Eudoxus the first proof that the volume of the cone is one-third of the volume of a cylinder with the same base and altitude, and much of Book XII is thought to be based on Eudoxus' work. Book XIII closes with a proof that there are only five regular Platonic solids, constructible from triangles, squares and pentagons. The solids are all constructed within a sphere, and there are also details on the relative distances from the edges of each solid to the centre. Here we find the irrationals described in Book X. The curtain then falls on a symphony in 13 movements.

The *Elements* has been the most influential textbook of all time. It has been copied and recopied, with commentaries overlaid upon earlier commentaries, translated and edited to fit the interests and culture of various civilizations. It has become almost impossible to reconstruct Euclid's original work, as only fragments have come down to us from before the ninth century AD, but it is a testament to the regard in which it was held that it has not only survived but has cast into oblivion all the other *Elements* before it.

Alexandria remained a centre of scholarship for some time. Apollonius (262–190 BC) of Perga, who often referred to it as the Great Geometer, studied and taught there. His most famous work is the advanced study of geometry, the *Conics*. The conic sections are the figures produced by slicing through a cone at various angles: the circle, ellipse, parabola and hyperbola. Archimedes studied there, as did Ptolemy and Diophantus (*c.* AD 250). The fourth century AD saw the tarnishing of Alexandria's freedoms. Hypatia (*c.* 370–415), daughter of Theon of Alexandria, the first woman mathematician to be recorded, became head of the Platonist school in Alexandria just as the increasingly powerful Christian movement was becoming less and less tolerant of what it perceived to be pagan science and philosophy. Her death at the hands of a local Christian sect is often seen as the beginning of the end for Alexandria as a centre of learning. As far as mathematics was concerned, the centre of gravity had shifted eastwards, to Baghdad.

chapter five

ten computational canons

◄ Frontispiece to a highly popular sixteenth-century text, the *Suanfa tongzong* (General Source of Computational Methods). The section entitled 'Discussions on difficult problems between Master and Pupil' shows the use of a counting board for mathematical calculations.

Chinese civilization first developed along the banks of the Changjiang (Yangtze) and Huang He (Yellow) rivers during the legendary Xia kingdom in the second millennium BC. The Shang Dynasty lasted from 1520 to 1030 BC, when it was supplanted by the Zhou invaders who, by the eighth century BC, began slowly to lose their grip on the territory. Then, from about 400 to 200 BC, what had become an empire disintegrated into a patchwork of feuding states. It is to this period, known as the Warring States, that we can trace the first purely mathematical text, the *Zhoubi suanjing* ('Zhou Dynasty Canon of Gnomonic Computations'). This was the time of Confucius, one of the many peripatetic scholars who lived a precarious existence advising local rulers. The re-unification of China under the emperor Qin saw both the rebuilding of the Great Wall and the wanton burning of books. In the succeeding Han Dynasty, from about 200 BC to AD 200, scholars searched out manuscripts which had evaded destruction, and often transcribed texts from memory. An influential mathematical text, the *Jiuzhang suanshu* ('Computational Prescriptions in Nine Chapters'), as well as commentaries on the 'Gnomonic Computations', belong to this period. The next major text appeared in the seventh century, when under the Sui Dynasty (518–617) and the Tang Dynasty (618–907) an educational reform programme led to mathematics being officially taught at the School for the Sons of the State. The textbook used was the *Suanjing shi shu* ('The Ten Computational Canons'), a compilation of the most important works then available, including both the 'Gnomonic Computations' and the 'Nine Chapters'. It remained influential for many centuries. The seventh century also saw the enormous engineering feat of linking China's two main rivers by the Grand Canal. The people eventually rebelled against the hardships imposed on them during its construction, and the short-lived Sui Dynasty quickly gave way to the Tang Dynasty. The Tang capital, Chang'an, developed as an intellectual bridge between China and Central Asia, playing a similar role to that other great cosmopolitan city far to the west, Baghdad. The three hundred years of Tang rule witnessed the invention of printing and of gunpowder. Our excursion ends with the Sung Dynasty, which lasted until the late thirteenth century. We shall now look at the 'Nine Chapters'.

Chinese interest in magic squares seems to be associated more with divination than with mathematics. Legend has it that Emperor Yu of the third millennium BC acquired two diagrams, one from a magical horse-dragon which rose out of the Huang He, the other from the shell of a turtle found in the Luo, a tributary of the Huang He. The first illustrations of the magical cross and square are from the tenth century, and no magic squares larger than 3×3 are discussed until the thirteenth century. By then the supposed magical properties were no longer being mentioned, and Yang Hui concentrates on the numerical properties of a variety of number squares and circles. In fact Arab mathematicians studied magic squares from the ninth century onwards, and an Arabic magic square dating from the Mongol era (1279–1368) has recently been found in Xi'an.

➤ The problem of the Broken Bamboo, from Yang Hui's *Xiangjie jiuzhang suanfa* (1261), a detailed commentary on the computational methods in the 'Nine Chapters'. The resulting right-angled triangle was used in a wide range of problems involving the Pythagorean theorem.

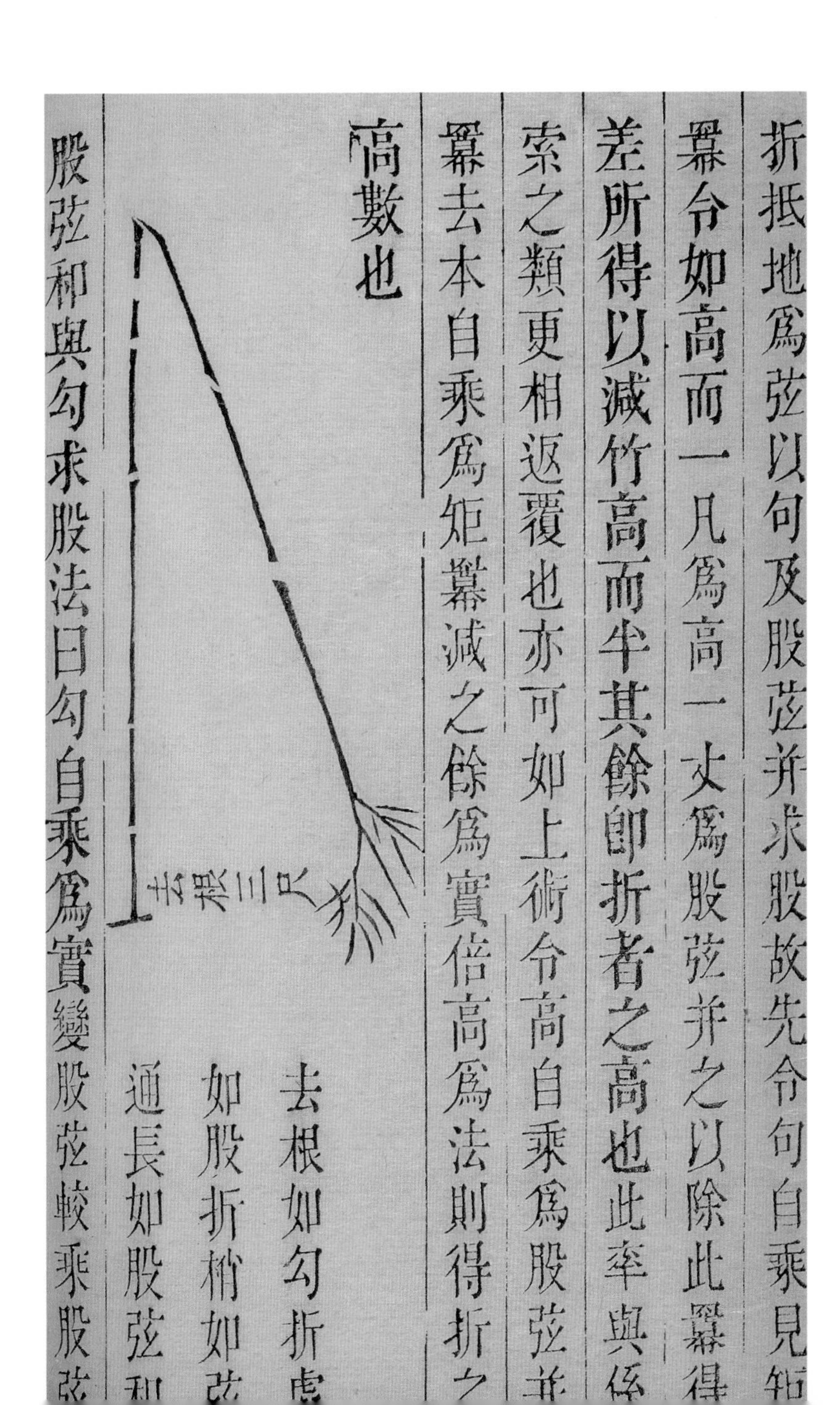
折抵地爲弦以句及股茲并求股故先令句自乘見
冪令如高而一凡爲高一丈爲股茲并之以除此冪得
差所得以減竹高而半其餘即折者之高也此率與
索之類更相返覆也亦可如上術令高自乘爲股茲并
冪去本自乘爲矩冪減之餘爲實倍高爲法則得折
高數也

股弦和與勾求股法曰勾自乘爲實變股茲較乘股

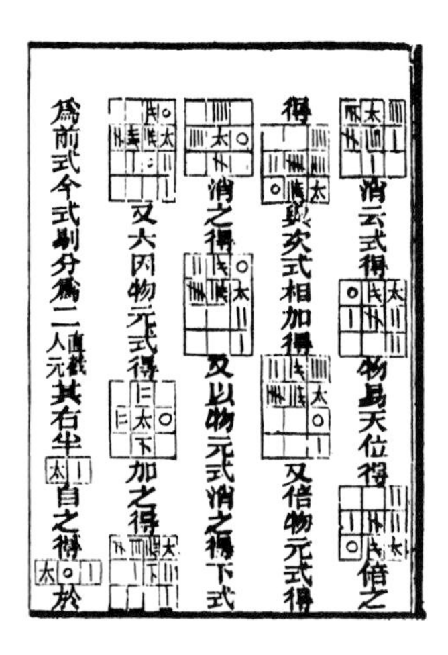

▲ A page from Shu Shijie's *Siyuan yujian* (1303), freely translated as 'The Jade Mirror of the Four Unknowns', illustrating the 'matrix' or 'array' notation used in seeking numerical solutions to algebraic problems.

The 'Nine Chapters' holds a key position in Chinese mathematics. The original is nearly impossible to disentangle from the mass of later commentaries; the third-century commentator Liu Hui states that the work was largely rewritten at the time, incorporating new material and discarding some unwanted sections. The earliest surviving version of the text dates from the thirteenth century, but this is only part of it; a more complete edition has come down to us from the eighteenth century. This is similar to our lack of original Greek texts, although here the timespan between existing artefacts and the originals they claim to reflect is rather longer. The 'Nine Chapters' contains 246 problems. Each begins with the problem stated, followed by the numerical answer and a method for reaching the solution. No logical explanations or proofs are given. A large part of the work consists of computational problems about practical tasks, such as apportioning land, dividing goods and managing large-scale building works. Here we shall look at methods for extracting square roots and for solving equations.

Calculations were performed by laying out counting-rods on a counting-board. Sometimes the counting-board was a special grid, but some texts mention that any surface could be used. The main thing is the arrangement of the rods during the calculation, which allows one to resume an interrupted calculation from where it was stopped, which is especially important in lengthy procedures. Answers would be recorded exactly as they appeared on the counting board. The resulting rod notation is a decimal place-value system, but the digits 1 to 9 are constructed using an additive system, with vertical rods for each unit and a horizontal rod to denote 5. Some sources have illustrations in which the directions of the rods change, but the 5 is always perpendicular to the units, which is undoubtedly a visual aid and speeds up calculations. The use of a special symbol for 5 is carried over to the abacus, which does not seem to have come into common use until the sixteenth century. Rather like the Babylonians, the Chinese did not appear to have a symbol for zero. The arrangements of counting-rods would have left a blank where a zero would have been, but this does not seem to have been transferred to answers when written down, and one could only distinguish from the context whether an answer was, say, 18, 108 or 1800. By the eighth century there is written evidence from a Chinese translation of an Indian text of a dot being used as zero. A circular zero first appears much later, in the thirteenth century, and also a 'square' zero, which is easily formed from counting-rods.

The extraction of square and cube roots starts by establishing the order of magnitude of the root from inspection, and each digit is then calculated in turn. In an example from the 'Nine Chapters', the square root of 71,824 is calculated. It is easy to see that the root lies between 200 and 300, and is therefore a three-digit number abc with a equal to 2. The object is then to calculate the values of b and c. The justification of the numerical procedure, as given by Liu Hui, stems from a geometric argument in which the square is dissected in a particular way. Having established that the root is in the 200's, we remove

▲ Liu Hui, the third-century commentator of the 'Nine Chapters', described a method of exhaustion to find an approximate value of π. This diagram by the scholar Dai Zhen (1724–1777) shows the method whereby a circle is approximated by inscribed polygons.

the square 200×200 from the diagram, leaving an L-shape called a 'gnomon'. We then find the highest possible value in the tens which fits into the gnomon. This we find to be 60, and another L-shaped gnomon is created. The process continues until the required solution is established. If the answer is not a whole number, either the process is continued to as many decimal places as desired, or a remainder is given as a fraction. The same technique is used to establish cube roots, by a similar dissection of a cube.

This geometric technique is equivalent to using the binomial expansion, whose numerical coefficients can be expressed by what is now known as Pascal's triangle. The explicit use of this algebraic method was certainly in place by the eleventh century, and possibly earlier, allowing the Chinese to calculate any *n*th root they required. Again, it is unclear whether Pascal's triangle was obtained from Indian sources or discovered independently. Each step in the extraction of a square root requires the solution of a quadratic equation. Similarly, the extraction of higher-order roots, such as a cube root, requires the solution of higher-order equations, or polynomials, such as a cubic. A similar method to root extraction could therefore be used to solve any polynomial, without the geometric framework of the gnomons. As in other cultures, the evaluation of a single root was always sufficient, and we can't tell whether the Chinese knew that a polynomial could have multiple solutions. The equations were not written in terms of a variable such as *x*, but were expressed in terms of just the numerical coefficients, which were laid out on the counting-board. They do not seem to have been concerned about whether the decimal places of the solution were finite or infinite – the algorithm was equally efficient in both cases, and the calculation would be terminated once a satisfactory accuracy was achieved.

The 'Nine Chapters' also includes problems which amount to systems of linear equations with more than one unknown. Liu Hui states in his commentary that the general method is difficult to explain without recourse to a specific example. In the method, the coefficients of the system of equations are represented by counting-rods laid out in an array, like a matrix. The numbers are then manipulated so as to eliminate some of the coefficients, leaving explicit numerical solutions. This is essentially identical to the modern method known as Gaussian elimination, after Carl Friedrich Gauss, but the Chinese did not go on to develop the idea of the determinant of a matrix, so it is possibly more correct to regard the arrangement as an array.

There is also important work on indeterminate equations, where there are a number of possible answers – sometimes an infinite number. Two types of problems are presented: the main one is the remainder problem, the other is known as the 'hundred fowls problem'. The hundred fowls problems occurs in a variety of guises across the medieval world in European, Arabic and Indian texts. As stated in the 'Ten Canons', cockerels cost 5 *qian*, hens 3 *qian* and 3 chickens 1 *qian*. If 100 birds are bought for 100 *qian*, how many of each bird can be bought? Three solutions are given, one of them is

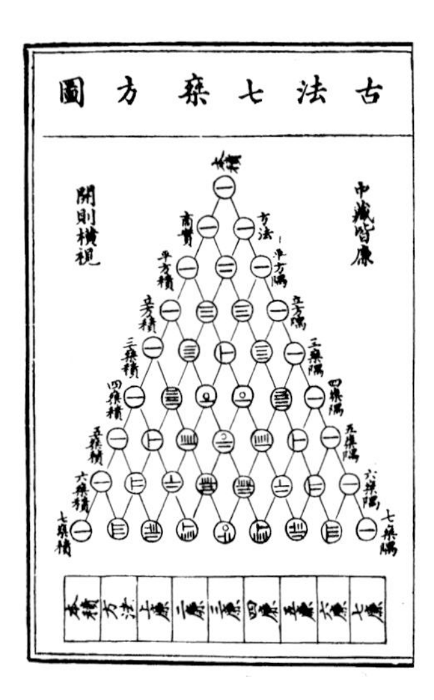

▲ The frontispiece of Zhu Shijie's *Siyuan yujian* (1303) shows what has become known as Pascal's triangle, a good three centuries before Pascal was even born.

4 cockerels, 18 hens and 78 chickens. (Missing is the solution where we have 25 hens and 75 chickens, but no cockerels.) The answers given are correct, but the explanation appears to be spurious.

The remainder problem, however, gives both the result and a general method, but yet again with no justification. The problem as stated in the 'Nine Chapters' assumes an unknown number of objects where, 'if they are counted in threes, 2 are left, if they are counted in fives, 3 are left, and if they are counted in sevens, 2 are left'. The aim is to find the number of objects. The solution as stated is procedural rather than explanatory; in essence the problem requires finding the highest common factor of the numbers 3, 5 and 7. Strangely, the next occurrence of these problems is not until the thirteenth century with the work of Qin Jiushao.

Born in Anyue, now in Sichuan, Qin Jiushao's father held a number of administrative positions, including Vice-Director of the Palace Library, Qin Jiushao studied at the Board of Astronomy in the capital, Hangzhou, but joined the war effort to repel the Mongol invaders in 1234, complaining of ten years of misery. He resurfaces in 1244 as a Court Gentleman for Comprehensive Duty in the prefecture of Jiankang, now Nanking, but later in the year retires from service for three years, mourning the death of his mother. It is probably during this period that he composed his *Shushu jiuzhang*, whose structure is similar to that of the 'Ten Canons' but far more sophisticated.

The *Shushu jiuzhang* describes the methods of solving an individual congruence and a set of simultaneous congruences such as the remainder problem. Congruences are perhaps better known in the form of modular or clock arithmetic, and the solutions given correspond to what is now known as the Chinese remainder theorem. Qin Jiushao states that he learnt the method from the calendar-makers working in the Bureau of Astronomy at Hangzhou, but that they used the rule without understanding it. The rule was introduced to resolve questions that arose through the use of different cycles, such as the lunar month, the solar year and the artificial sexagesimal cycle. In fact even Gauss, who rediscovered the method five centuries later, used problems of calendar cycles as examples. It is not clear where Qin Jiushao actually obtained his rule. It seems to have been a genuine innovation by a first-rate mathematician, going beyond the tradition of commentaries. He did, however, exemplify the long-standing Chinese tradition of computational innovations in the service of real-life problems.

chapter six

mathematical sutras

◀ Detail of the astrologers using an astrolabe and tables of planetary positions in casting the horoscope at the birth of Tamerlane (1336–1405), the future Mongol emperor.

The earliest known evidence of mathematics in Asia comes from the Harappan civilization of the Indus Valley around 3000 BC. The earliest documents, though difficult to decipher, seem to deal largely with trading accounts, and weights and measures, with special reference to an advanced brick-making technology. In about 1500 BC the Harappan culture was destroyed by invaders from the north. These so-called Aryans were a pastoral people speaking an Indo-European language, the forerunner of Sanskrit and many of the world's modern languages. The first recorded codification of any language was set down by the great grammarian Panini during the fourth century BC, who single-handedly made Sanskrit a robust and subtle language capable of encoding the thoughts of the subcontinent for over two thousand years. If Greek mathematics can be said to have arisen out of philosophy, then Indian mathematics has its roots in linguistics.

The earliest Vedic literature is mainly religious and ceremonial, the most important for their mathematical content are the appendices to the main Vedas, known as Vedangas. These are set down as sutras – short poetic aphorisms, peculiar to Sanskrit writings, which strive to give the essence of an argument in the most condensed and memorable form. The Vedangas are classified into six fields: phonetics, grammar, etymology, verse, astronomy and rituals. It is the last two subjects that provide us with insights into the mathematics of the time. The *Vedanga* on astronomy is called *Jyotisutra*, while that on rules for rituals is known as the *Kalpasutras*, a part of which deals with the construction of sacrificial altars, the *Sulbasutras*.

The earliest *Sulbasutras* were written in about 800–600 BC, before Panini's codification of Sanskrit. The geometry grew out of the need to conform to the size, shape and orientation of altars laid down in Vedic scriptures. Absolute accuracy was as essential for the efficacy of the ritual as was the pronunciation of the mantras. The geometry is expressed in essentially three ways: explicitly stated geometric theorems; procedures for constructing various shapes of altar; and algorithms related to the previous two categories. The most important theorem stated is the Pythagorean theorem for right angled-triangles.

One example will illustrate how theoretical results went hand in hand with practical concerns. Using the Pythagorean theorem, it is always possible to construct a square whose area is twice that of any given square. But if we start with two real squares – made from cloth, say – what is the most efficient way of cutting them up and recombining the pieces to form the larger square? Although this type of construction is not explicitly stated in the *Sulbasutras*, there is evidence of such concrete modes of thinking. One clue is the approximation used for $\sqrt{2}$, which is accurate to five decimal places: 'Increase the measure by its third and this third by its own fourth less the thirty-fourth part of that fourth.' This could represent cutting one of the squares into suitable rectangles and arranging them around the other square to construct a square of twice the area. This approach has similarities to Chinese geometry, and the value is very close to that found by the Babylonians.

➤ A scene from the *Akbar Nama*, a late sixteenth-century visual chronicle of Mughal India, depicting the birth of Tamerlane, the Mongol emperor whose descendant later founded the Mughal empire.

Given the prominence of the Hindu-Arabic numerals in the place-value decimal system, it is worth looking briefly at the early history of Indian numerals. Kharosthi numerals are found in inscriptions from the fourth century BC. There are special symbols for one and four, and for ten and twenty; numbers up to one hundred are built up additively. The earliest traces of Brahmi numerals are found in the third century BC, on the Asoka pillars scattered around India, and were more developed, including special symbols for multiples of ten and of one hundred, as well as for higher powers of ten. The Bakhshali numbers are of very uncertain antiquity, but if dated correctly to the third century AD they would be the first known place-value system with a special symbol for zero. With just ten symbols, it was thus possible to express any number, however large. The Gwalior numerals of the ninth century AD are recognizably similar to our modern numerals, and the first undisputed occurrence of the zero in an Indian inscription. Outside India, but still within its cultural influence, we find a Khmer inscription in Cambodia dated to AD 683 which shows the use of zero.

The classical period of Indian mathematics began in the middle of the first millennium. Much of India was ruled by the Imperial Guptas, who encouraged studies in the sciences and arts. Mathematical activity was concentrated in three centres: Kusum Pura, the imperial capital, Ujjain in the north and Mysore in the south. The two most important mathematicians of this period are Aryabhata (476–550), author of the *Aryabhatiya*, and *Brahmagupta* (598–670), who in 628 wrote the *Brahmasphutasiddhanta* ('The Opening of the Universe'). The main concerns of these two were mathematical astronomy and the analysis of equations.

The *Aryabhatiya* is written in 33 verses, beginning with a benediction, and proceeds to present algorithms for calculating squares, cubes, square roots and cube roots; 17 verses then deal with geometry, and 11 with arithmetic and algebra. The tenth verse gives a value of π as the ratio 62,832 : 20,000, equivalent to 3.1416, the most accurate value that would be obtained for nearly a thousand years. The work also includes a table of sines. In contrast to Ptolemy's use of the chord as the basic measure, Indians used the half-chord and expressed it in terms of the radius. Therefore, except for a constant factor, the Indian sines are close to our current concept. Dividing the quadrant into 24 equal parts, and starting from some basic results and formulas, such as $\sin 30° = ½$, Aryabhata computed a table of sines for angles from $3° 45'$. He is also credited with a formula for approximating the sine of any angle without the use of a table, which is generally accurate to a couple of decimal places. Later, Brahmagupta gave an interpolation formula using an arithmetical method of differences to find the sines of intermediate angles. Trigonometry would be further refined by the Arabs in the north and by the Kerala mathematicians in the south. It was partly through a translation of the *Brahmasphutasiddhanta* that the Arabs and then the West became aware of Indian mathematics and astronomy.

▲ Astronomers observing stars using a theodolite, able to measure both vertical and horizontal distances, and consulting the Sanskrit texts on astronomy and trigonometry known as the *Siddhantas*.

Brahmagupta is one of the best known of the Ujjain school. His *Brahmasphutasiddhanta* is a comprehensive treatise of the astronomical knowledge of the time. Some of the mathematical sections deal with indeterminate analysis, which occurs in calendrical calculations and astronomy. Aryabhata solved linear indeterminate equations, using the Euclidean algorithm described in the *Elements* to reduce the size of the coefficients until the equations can comfortably be solved by trial and error. Brahmagupta gives an algorithm for whole-number solutions of equations of the form $ax^2 \pm c = y^2$, which geometrically are hyperbolas, and in Europe the relation came to be known as Pell's equation. These methods were refined by Bhaskaracharya into a 'cyclic' method known as *chakravala*. He gives a solution to a famous problem, the equation $61x^2 + 1 = y^2$. This is precisely the problem that Pierre Fermat set as a challenge in the seventeenth century, a solution only being found by Joseph-Louis Lagrange a hundred years later; even then, the chakravala algorithm is far more efficient. The smallest solutions are $x = 226{,}153{,}980$ and $y = 1{,}766{,}319{,}049$.

Neither the *Aryabhatiya* nor the *Brahmasphutasiddhanta* prove the results they present. But this does not mean that their authors were unaware of their proofs, or of the need to demonstrate the validity of the rules given. The importance of proving results can be traced back to Bhaskara, who rejected the Jain approximation to π of $\sqrt{10}$ because, though numerically close, there was no meaningful derivation. Thus the mere presentation of results and procedures is extended by the verification of those results, which in its turn gives way to more rigorous derivations.

Bhaskaracharya (1114–1185) was the most distinguished mathematician from Ujjain, and is credited with some concepts that would much later appear in the development of the calculus. His manuscripts were still being published in the nineteenth century. One aspect of Indian astronomy was the study of the instantaneous motion of planets, and especially of the Moon. Very accurate timings of eclipses were made so that future eclipses could be predicted with pinpoint accuracy. Both Aryabhata and Brahmagupta use a formula for this, and Bhaskaracharya extended the result to give what appears to be the differential of the sine. In the *Siddhantasiromani* there is an 'infinitesimal' unit of measure, the *truti*, equal to 1/33,750 of a second. This 'pre-calculus' was confined to astronomical work, and does not appear to have been considered as a general topic on its own or transferred to other branches of mathematics.

In his formulation of the calculus, Newton made great use of infinite series. Especially useful was the approximation of sines and cosines by appropriate polynomials with an infinite number of terms, and we find specifically in Kerala developments in just this direction. After Bhaskaracharya, Indian mathematics made little progress and the country was in political turmoil. But south-western India remained largely sheltered from these upheavals, and developed its mathematics further between the fourteenth and seventeenth centuries. Kerala was a centre of maritime trade and was a cosmopolitan environment. The history of Kerala's role in the traffic of ideas has yet to be established, but a few mathematical results point to a thriving indigenous mathematics.

Madhava of Sangaṃagramma (*c.* 1340–1425), known to later astronomers as Golavid, or 'Master of the Spherics', was one of the greatest medieval mathematicians. His works on infinite series have been lost, but are quoted extensively by later writers in the sixteenth century. Many results which have been named after European mathematicians may need to be suffixed by the name of Madhava. These include infinite polynomial expansions of sines and cosines, which have been credited to Newton, as well as small-angle approximation formulas, which are part of the general Taylor series. These would have allowed trigonometry tables to be drawn up to any desired accuracy; Madhava's tables were accurate to eight decimal places. We also find various infinite series expressing the value of π. One, given in verse, illustrates how certain objects were traditionally used to signify numbers to aid recollection:

Gods [33], eyes [2], elephants [8], snakes [8], fires [3], three [3], qualities [3], *vedas* [4], *naksatras* [27], elephants [8], and arms [2] – the wise say that this is the measure of the circumference when the diameter of the circle is 900,000,000,000.

Reading the numbers from right to left and dividing by the diameter yields a value of π accurate to eleven decimal places. Such a facility with handling infinite series recalls a modern prodigy from Kerala, Srinivasa Ramanujan (1887–1920), whose incredible results propelled him to Cambridge University.

chapter seven

the house of wisdom

◄ One of the earliest such instruments, this Arab astrolabe was constructed by Ahmad ibn Khalaf in ninth-century Iraq. The astrolabe is a kind of analogue computer and can be used to measure time and predict the positions of celestial bodies, as well as to survey.

In the seventh century AD the Arabian peninsula gave birth to a new monotheistic religion which was to spread across the Christian and Persian worlds. In AD 622 the prophet Muhammad fled from Mecca and sought sanctuary in Medina. Only eight years later he returned at the head of an army and entered Mecca in triumph. Inspired by Muhammad's revelations, his followers spread the message of the Koran and established an Islamic empire which at its height spread from Cordoba to Samarkand. The early empire was ruled by the Umayyad dynasty, with their capital in Damascus. But in 750 they were overthrown by the Abassids, who moved their capital to Baghdad, the Umayyads fleeing to the Spanish territories to set up a rump caliphate.

The Abassid caliphs sought to build the new Alexandria in Baghdad and founded an astronomical observatory, a library and a research centre called Bait al-Hikma ('The House of Wisdom'). A massive translation project was undertaken to render into Arabic all the precious knowledge available at the time. In Arab mathematical sciences we can see the influences of Babylonian, Indian and Greek ideas. Their synthesis and development led to fundamental work, especially in algebra and trigonometry. Although algebraic symbolism as we would recognize it today was a much later European development, algebraic thinking can probably be ascribed to Arabic mathematicians. Earlier mathematics can often be interpreted algebraically, but the explicit recognition that geometric problems can be expressed algebraically, that geometric procedures can be translated into algebraic algorithms, that algebraic procedures can be extended beyond their geometric roots – these are all Arab contributions.

An influential work in the history of algebra is the *Arithmetica* of Diophantus of Alexandria (*c.* 200–*c.* 284). It is still uncertain in which century to place Diophantus, although solving a mathematical conundrum reportedly inscribed on his grave gives his age at his death. The *Arithmetica* is seen as a new strand in Greek mathematics, with its focus on solving determinate and indeterminate equations numerically, without reference to geometric justifications. The restriction to whole-number solutions is now a branch of mathematics known as Diophantine equations, an example of which is the search for Pythagorean triples. Diophantus also used what is called a syncopated algebraic notation, an intermediate stage between a rhetorical and a wholly symbolic algebra. This work was translated and widely studied by the Arab mathematicians.

Abu Jafar Muhammad ibn Musa al-Khwarizmi (*c.* 780–*c.* 850), to give him his full name, is one of the most important Arab mathematicians. His name suggests that he came from Khwarizm, in Central Asia. He seems to have spent most of his life in Baghdad, and was made chief librarian of the newly founded House of Wisdom. His treatise on algebra, *Hisab al-jabr w'al-muqabala*, or 'Calculation by Restoration and Reduction', would later be highly influential in Europe; indeed, our word 'algebra' comes from a Latin transliteration of *al-jabr*. Al-Khwarizmi's motivation was to solve practical problems in trade, inheritance and land use. The sections on algebra cover linear and quadratic

➤ A sixteenth-century Turkish manuscript by Loqman, the *Zubdat al-Tavariq* ('Treasure of History') illustrates the esoteric side of Muslim cosmology. Each of the 'planets' corresponds to a prophet, including Moses and Jesus. Beyond the zodiac and lunar mansions we see the angelic realm, gateway to the Divine Presence, and appearing to turn the cosmos.

▲ *Takiyuddin in His Observatory at Galata*, an illustration from *Shahinshahnama* ('Book of the King of Kings') by Loqman in the sixteenth century. Note the range of mathematical and astronomical instruments in use, including an astrolabe, quadrants, set square and compass, and at the upper left a diopter. Founded in 1575 the observatory was short-lived as its astrological predictions proved unpopular.

equations – the terms 'restoration' and 'reduction' referring to algebraic manipulations. Al-Khwarizmi classifies quadratics into six different types. Rather than writing a general quadratic as $ax^2 + bx + c = 0$ with x as the unknown and a, b and c the coefficients, his algebra requires all coefficients and all answers to be positive. The way the equation is arranged here would have been meaningless to him because the sum of positive terms could never be equal to zero, and he would treat the equations $ax^2 + bx = c$ and $ax^2 + c = bx$ as two different types. Algebraic solutions are given for each type followed by a geometric demonstration, possibly employing results from Euclid but also bearing similarities with Babylonian and Indian methods. The geometric demonstrations of algebraic methods are still rhetorical: al-Khwarizmi did not develop a symbolic language, but the ease with which we shift between the realms of algebra and geometry seems very different to the Greek style.

By the time of al-Karaji (953–*c.* 1029), Arab mathematicians were attempting to free algebra from geometric thinking and make it into a general technique for handling unknowns arithmetically. Al-Karaji founded a highly influential school of algebra in Baghdad. His major work on the subject is the *al-Fakhri*, in which he defines higher

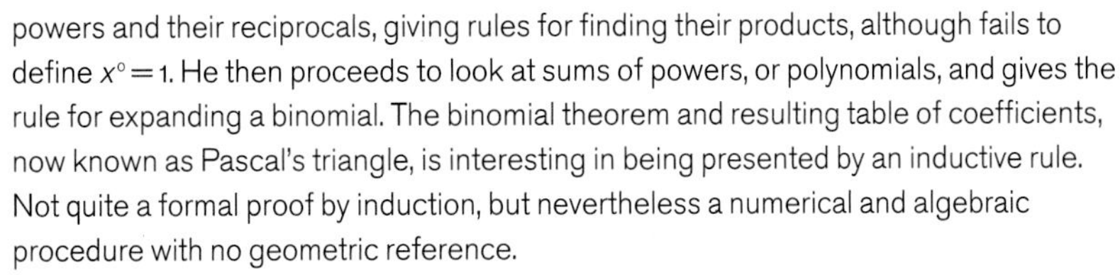

powers and their reciprocals, giving rules for finding their products, although fails to define $x^0 = 1$. He then proceeds to look at sums of powers, or polynomials, and gives the rule for expanding a binomial. The binomial theorem and resulting table of coefficients, now known as Pascal's triangle, is interesting in being presented by an inductive rule. Not quite a formal proof by induction, but nevertheless a numerical and algebraic procedure with no geometric reference.

By the time of Ghiyath al-Din Abu'l-Fath Umar ibn Ibrahim Al-Nisaburi al-Khayyami, better known as Omar Khayyam (1048–1131), the Seljuk Turks had taken Baghdad and proclaimed an orthodox Muslim sultanate. After studying at Nishapur, in 1070 Khayyam left this inflammable political situation for the relative calm of Samarkand, in modern Uzbekistan. Although better known as a poet and writer of the *Rubaiyat*, he was chiefly a scientist and philosopher. It was at Samarkand that he wrote his *Algebra*, the most original part of which was the solution of cubic equations by geometric means. His insight was that solutions to cubics could be found from the points of intersection of two conic sections, with which he was familiar from a translation of Apollonius. For example, an equation of the form $x^3 + ax = c$ could be solved as the intersection of a suitably constructed circle and a parabola. He classified certain cubics and their solutions, giving algebraic methods for simplifying some complicated cubics into known types or into the even simpler quadratics. Although this may seem a step backwards in terms of algebraic developments, there are a number of aspects which make Khayyam's contribution unique. He comments that the ancients left nothing on the solution of the cubic, and we must assume he had ample access to the best libraries in the land. He also states that a geometric solution to the cubic cannot be found by ruler-and-compass methods, a proof of which was not discovered for another seven hundred years. He was the first to realize that cubics may have more than one solution, but did not appreciate that there could be three. Khayyam recognizes that his work has not been completed, and looks forward to a wholly algebraic solution to cubics and higher-order equations analogous to the formula for quadratics. This had to wait for the Italian Renaissance. Khayyam's analytic geometry was the high point of the Arabic fusion of algebraic and geometric knowledge. The next major step was not taken until the work of Descartes.

Astronomy was a major interest of Arab mathematicians, and their developments in trigonometry made it possible to construct ever more accurate astronomical tables. Islamic religious ritual had a galvanizing effect on mathematics as an aid to ensuring that the rules were accurately followed. The Islamic calendar was based on lunar months, each month beginning on the first appearance of the lunar crescent after new Moon. The five daily prayers had to be performed at times regulated by the Sun's position: for example, the afternoon prayer has to take place when the length of an object's shadow cast at noon has increased by an amount equal to the height of the object itself. And the faithful had to pray facing in the direction of the Kaaba in Mecca. All three of these rules

▼ Nasir al-Din al-Tusi (1201–1274) at the observatory he founded in Maragheh in present-day Azerbaijan. Persian and Chinese astronomers collaborated at the observatory, which boasted a wall quadrant 4 metres in length and a very fine library. After 12 years of observations al-Tusi published his *Ilkhanic Tables* of planetary and stellar positions.

demanded knowledge of celestial and planetary motions, as well as terrestrial geography. Initially there were observational methods of keeping approximately to the required rules, and there were tables from Greek and Indian sources. The Arabs greatly refined both the tables and the methods of observation, and by the thirteenth-century mosques employed professional astronomers who were skilled in the use of astrolabes, quadrants and sundials.

It became obvious that any advances in astronomical calculations demanded accurate trigonometric tables. Let us look at these advances through the methods used to find the sine of 1 degree. The sine, cosine and tangent had all been suitably defined, and a variety of formulas, such as for the sine of the sum and difference of two angles, were well known. The general technique was to start with those sines which were accurately known from geometric calculations, such as $\sin 60° = \sqrt{3}/2$ or $\sin 30° = 1/2$, then use the half-angle formulas to continually halve the angle until one gets to 1°, or close to it. Abu-l-Wafa (940–998) started with the known value of sin60° and also calculated sin72°, and from the appropriate formula he was thus able to calculate sin12°. Using the half-angle formula he progressed down to sin(1° 30′) and sin45′. As these two angles are very close to each other, he assumed that the intermediate values are in an approximately linear relationship, and an arithmetic method would thus yield the required value of sin1°. By using techniques such as this, Abu-l-Wafa was able to construct a complete table with angles to the nearest ¼°, or 15′ in sexagesimal. He achieved an accuracy of 5 sexagesimal places, or 8 decimal places.

I was unable to devote myself to the learning of this algebra and the continued concentration upon it, because of obstacles in the vagaries of time which hindered me; for we have been deprived of all the people of knowledge save for a group, small in number, with many troubles, whose concern in life is to snatch the opportunity, when time is asleep, to devote themselves meanwhile to the investigation and perfection of a science...

Omar Khayyam, *Treatise on Demonstration of Problems of Algebra*, *c.* 1070

The next major step was not taken for another three hundred years, even though the theory was readily available. By that time Baghdad was under Mongol rule and the emperor Ulugh Beg (1394–1449) was building his scientific centre at Samarkand. Al-Kashi (1380–1429), the first director of Samarkand's new observatory, improved dramatically the accuracy of the sine tables. Using the triple-angle formula for sines, he set up a cubic equation to find sin 1° in terms of sin 3°. Then, using an iterative procedure, he calculated sin 1° to 9 sexagesimal places, equivalent to 16 decimal places. The rest of the table could then be completed with established relationships, but this was still a phenomenal feat of calculation. A similar method was to be used by Johannes Kepler two hundred years later. Alongside increased numerical accuracy, the Arabs perfected the astrolabe as both an instrument for observation and as an analogue calculator that used the heavens to tell the time. But Baghdad's star was waning. The Mongol invasion was followed by that of the Ottoman Turks, whose capital and intellectual centre was established in Istanbul.

chapter eight

the liberal arts

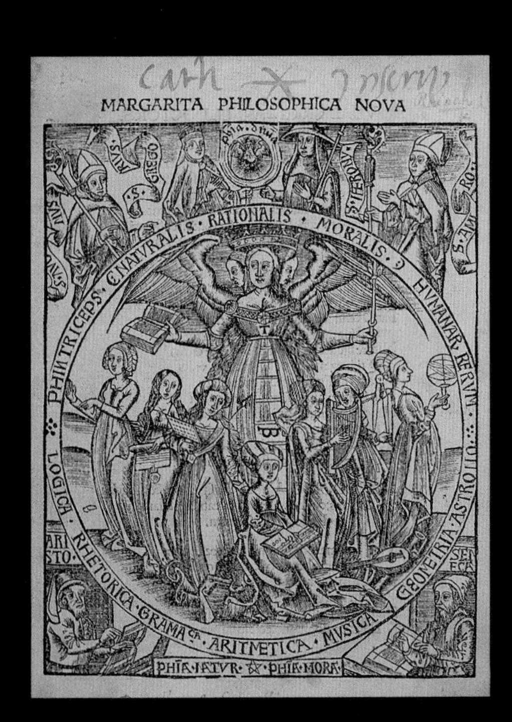

◄ Frontispiece from Gregor Reisch's *Margarita Philosophica* (1503), showing all the seven liberal arts: logic, rhetoric, grammar, arithmetic, music, geometry and astronomy. The two figures at the bottom are Aristotle and Seneca.

In AD 529 the Christian Roman Emperor Justinian closed the pagan philosophical schools, including the Academy in Athens. A thousand years of Greek mathematics thus came to an end, and many scholars made their way eastward to the intellectually more fertile Persian Empire. Two hundred years before, Constantine the Great had made Christianity the official religion of the Roman world, and moved the seat of power from Rome to Byzantium, which he renamed Constantinople. The spiritual and temporal powers were briefly united in the first Holy Roman Emperor, Charlemagne (742–814). In his time, Constantinople was part of the emerging Islamic empire, and Baghdad was the scientific capital of the known world. As ruler of the western European empire Charlemagne, was concerned at Christendom's intellectual inferiority and instigated educational reforms centred on cathedral schools. This was the responsibility of Alcuin of York (735–804), head of Charlemagne's Palace School at Aachen. Alcuin also developed the Carolingian minuscule script, which is the basis of the modern Roman lower-case letter-forms. On Charlemagne's death his three squabbling sons once again divided Europe. Education was far from their primary concern, but a rivulet of scientific learning was preserved in the religious schools and monasteries.

The curriculum of the seven liberal arts had been laid down during Roman times. It was divided into the trivium of grammar, rhetoric and logic, and the quadrivium of geometry, arithmetic, astronomy and music. Mathematics may thus appear to have been a key part of the curriculum, but in reality the level of understanding was elementary. Boethius (*c.* 480–524), probably the foremost mathematician the Roman world produced, laid down what became the standard texts for each branch of the quadrivium. His *Arithmetic* was simply an abridgement of a late Alexandrian work, the *Introduction to Arithmetic* by the noted Pythagorean Nichomacus (*c.* AD 60–*c.* 120); the *Geometry* was based on Euclid's first four books, minus the proofs; the *Astronomy* was a stripped-down version of Ptolemy's *Almagest*; and the *Music* was a compendium from Greek sources. The syllabus seemed to be designed to maintain minimum standards rather than provide a springboard to new discoveries. The mathematics was mainly used to maintain the calendar and calculate the date of Easter, both of which required knowledge of astronomy. The scientific reawakening of Latin Europe was sown by the remarkable cross-fertilization of ideas taking place along the borders of the Christian and Islamic worlds.

Inspired by the prophet Muhammad and the teachings of the Koran, the Arabs exploded out of their peninsula to conquer the Persian and eastern Roman empires. The borders with Latin Europe extended from southern Spain and Sicily to the eastern provinces. It was in Spain, and especially in the city of Toledo, that an intellectual dialogue was conducted between two cultures which were at the same time in near-perpetual conflict with each other. It is almost a miracle that such a climate of academic tolerance was achieved during a period that saw two centuries of crusades. Toledo had been the capital of the Visigoths before it was invaded by Arab forces in the eighth

➤ *Representation of Astronomy* from Gregor Reisch's *Margarita Philosophica* (1503). The figure is holding a quadrant which, with the aid of astronomical tables, could be used to measure one's latitude and the time of day.

▲ A fourteenth-century astronomer using an horologium cum fistula, a sighting tube aimed at the Pole Star, allowing nocturnal time-keeping.

◄ *Representation of Geometry* from Gregor Reisch's *Margarita Philosophica* (1503). The illustration shows the very practical nature of geometry, from the construction of a quadrant through to surveying, carpentry and architecture.

century, only to be retaken by Christian armies at the close of the eleventh century. Cordoba became the capital of the Iberian Arab state, and its Umayyad rulers had plans for it to surpass Abbasid-governed Baghdad in splendour and learning. The sultanate of Granada, as the last stronghold of Islamic Spain was known, was to persist until 1492, when Muslims and Jews were expelled from Catholic Spain. This western outpost of the Arab empire achieved its aim of matching Baghdad as a haven for the arts and sciences, Christians, Muslims and Jews working together to establish a corpus of important works in all the major languages. Translations were undertaken between Arabic, Latin, Greek, Hebrew and Castilian. For Europe this was a crucial period in its rediscovery of lost Greek mathematics and original Arabic and Indian contributions. The cosmopolitan character of eleventh- and twelfth-century Toledo can be seen from the names of some of the leading scholars of the period: Robert of Chester, Michael Scot, Hermann of Carinthia, Plato of Tivoli, Eugenio of Palermo, Rudolph of Bruges, John of Seville, Gerard of Cremona and Adelard of Bath.

Adelard of Bath (1075–1160) is probably the most famous translator who is conspicuously absent from Toledo's roll of honour, and we assume that his knowledge of Arabic was gained in Sicily, which had changed from Arab to Norman rule a century earlier but had preserved the Islamic spirit of learning. He translated al-Khwarizmi's astronomical tables in 1126 and Euclid's *Elements* in 1142, both from Arabic into Latin, and Ptolemy's *Almagest* from Greek into Latin in about 1155. Little else is known about Adelard's life except that he travelled widely in France, Italy and Turkey.

Perhaps the greatest translator was Gerard of Cremona (1114–1187), to whom have been ascribed over 85 translations. He went to Toledo originally to learn Arabic in order to read Ptolemy's *Almagest*, which at the time did not exist in Latin. He remained there for the rest of his life, translating works on mathematics, science and medicine. Among them was a revised version of Thabit ibn Qurra's Arabic version of Euclid's *Elements*, an improvement on Adelard's earlier work. The first translation of al-Khwarizmi's *Algebra* in 1145 was made by Robert of Chester. It is during this period that many words which are now common entered the European lexicon, often through misunderstandings of the technical language or poor transliteration. Words such as 'algorithm' and 'algebra' were corruptions of 'al-Khwarizmi' and *al-jabr* from the full title of his *Algebra*, *Hisab al-jabr w'al-muqabala*. In the full Arabic title the term al-jabr means 'completion', and refers specifically to the method of removing negative terms from an equation. Other words, among them 'nadir', 'zenith', 'zero' and 'cipher', come from this period.

It was not long before these translations inspired a quest for new knowledge. Early Church doctrines had assimilated a fair dose of Platonic philosophy, and yet in 529 AD the emperor Justinian closed the Platonic Academy in Athens, nine hundred years after its founding, for fear of such pagan philosophies. At about the same time, Aristotelian logic had been enshrined within Boethius's trivium. Both Plato and Aristotle were, in

▲ Richard of Wallingford (*c.*1292–1336), Oxford mathematician and astronomer, later became Abbot of St Albans. Shown here constructing an instrument, possibly an astrolabe, with a pair of compasses.

◄ A celestial chart showing the Signs of the Zodiac, from the Catalan Atlas by Abraham Cresques in the fourteenth century.

their different ways, closely bound to Christian theology. The critical re-evaluation of Greek science and philosophy was thus seen in some quarters as another attack on the authority of the Church itself. Aristotle had written on a wide range of scientific topics including mechanics, optics and biology. Unfortunately, despite laying emphasis on observation, many of his theories seemed to contradict direct experience. Plato, on the other hand, wrote relatively little about science, and was often scornful of its practice, but he did stress the primacy of mathematics in describing the universe. To Aristotle, mathematics was subordinate to physics. The situation was further complicated by the translations of Arabic and Greek works which contradicted each other. The pre-eminent centres of study at the time were Paris and Oxford, and we shall look in particular at the movement known as the Merton School, centred around Merton College, Oxford. The nascent scientific method was to see mathematics play a central role.

This new philosophy of rational enquiry was initiated by Robert Grosseteste (1168–1253). Educated at Merton College, he was Chancellor of the University from 1215 to 1221, lecturer to the Franciscans in Oxford between 1229 and 1235, and then Bishop of Lincoln, in which diocese lay Oxford. Mathematics on its own is largely theologically neutral, but the combination of mathematics and physics presented a powerful challenge to established cosmological doctrines. The medieval science of optics illustrates this well. Grosseteste shows some Neoplatonist sympathies in the importance he ascribes to light as being the foundation of the whole universe. He had a cosmological theory, reminiscent of our Big Bang, in which the universe began as a flash of light and condensed into matter as it expanded. He generally followed the Arab writers, such as al-Haytham (perhaps better known by the Latinized version of his name, Alhazen), in preference to the Greeks, such as Aristotle. He claimed that light was a material pulse propagating through the air in a straight line, with similarities with the propagation of

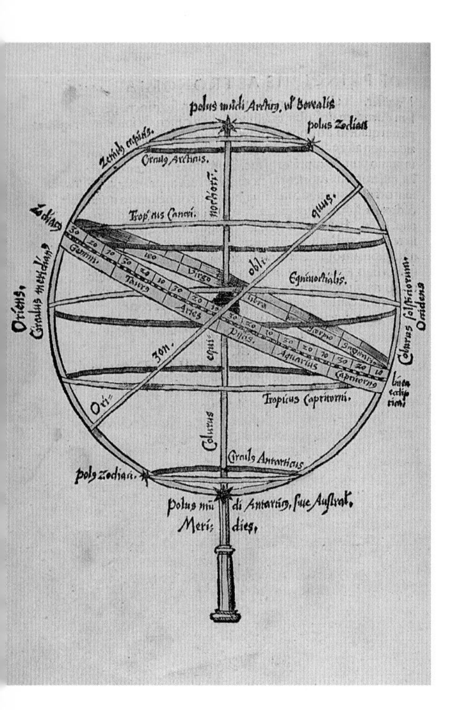

▲ An armillary sphere from the book of Astronomy in Gregor Reisch's *Margarita Philosophica* (1503). Used mainly for teaching, it shows the Earth at the centre and the key circles such as the ecliptic on the celestial sphere.

sound. Both travelled at a constant speed, but it was clear that light travelled faster. He experimented on lenses and described their use in magnifying objects. Arabs were making lenses in the eleventh century, and in thirteenth-century northern Italy spectacles were being manufactured, although of variable quality. Grosseteste thought that a rainbow was produced by a cloud acting as a lens, light being doubly refracted as it entered and then exited it, in contrast to Aristotle, who thought it was produced by light reflected from water droplets. Grosseteste's leading student, Roger Bacon (1214–1294), took this further, analyzing the apparent centre of a rainbow, its diameter and its spatial relationship with the Sun and the observer. He too held that the rainbow was produced by internal refraction, in each individual droplet rather than the whole cloud. The writings of Bacon, who was known as Doctor Mirabilis in his time, cover a vast swathe of mathematics and science. His speculations on submarines and aeroplanes can be compared with the later Leonardo da Vinci (1452–1519). At the end of the thirteenth century, the German writer Theoderic of Freiberg (d *c.* 1311) experimented with spherical glass flasks filled with water and with crystal balls to simulate water droplets. His observations led him to the theory of internal refraction of light and the splitting of the colours within the body of the droplet or glass. This theory is now usually ascribed to René Descartes, but we can see that three hundred years earlier the medieval scientists were making similar progress in the science of optics.

In some quarters, the star of Aristotelianism was beginning to wane. Roger Bacon wrote that, 'If I had the power over the works of Aristotle, I would have them all burned.' He saw them as retarding progress through their overreliance on philosophical dogma rather than empirical observation. His forthright ideas landed him in prison, as happened to other intellectuals of the age. William of Ockham (*c.* 1288–1349) continued the assault on Aristotle, maintaining that theology and natural philosophy should be separated, one dealing with knowledge from revelation, the other from experience. What is now known as 'Ockham's Razor', which had already been stated by Grosseteste, is the philosophy that in science one should look for the simplest solution that fits the facts. The charge levelled at both theology and scholastic philosophy was that they sought to explain physical reality through a deductive system based on absolute assumptions. What the medieval scientists sought was an inductive step from experimental data to a physical hypothesis which, expressed in the language of mathematics, would allow verifiable consequences to be deduced. We can see that these medieval scientists made a supreme effort to create a workable empirical philosophy.

William of Ockham's premature death in 1349 was caused by the Black Death that rampaged through Europe. It is not clear whether the plague was wholly to blame for the dip in mathematics and science, or whether the religious revival in its aftermath subdued any rebellious spirits. Whatever the reason, the medieval scientific spirit was nipped in the bud, and it would be another two hundred years before it flowered again.

chapter nine

the renaissance perspective

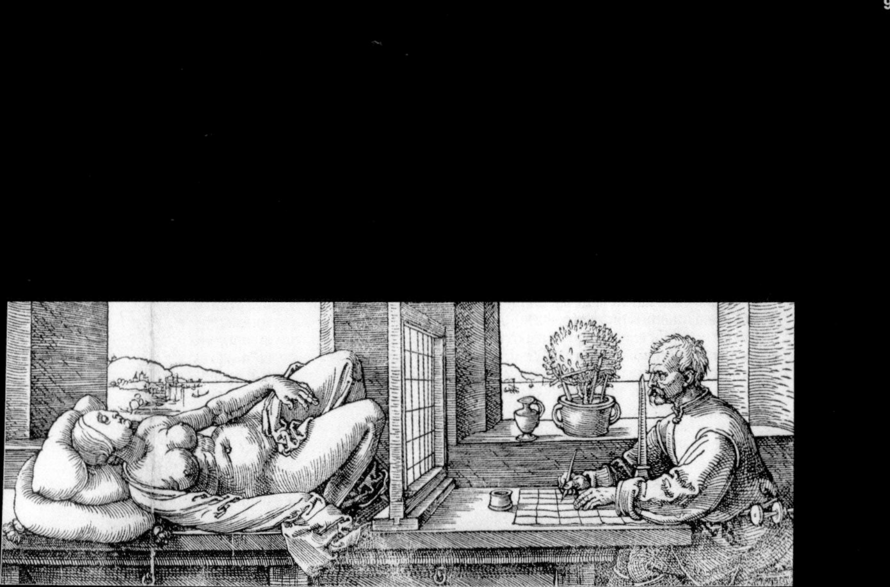

◄ Albrecht Dürer, *Treatise on measurement with compasses and straightedge* (Nuremberg, 1525), showing a veil and grid used to draw an image in perspective.

Much has been written about the Italian Renaissance as the defining period in a new European consciousness. The reawakening to classical learning was allied with a desire to go beyond mere mimicry and explore new styles, new ideas and new lines of investigation. The interactions between art and geometry, and specifically the use of perspective, illustrate these new directions well. The naturalistic style characteristic of the Renaissance was apparent in the arts before the study of perspective bore fruit, but perspective injected an added realism by formally incorporating the viewer's point of view in the fabric of a painting. Perspective was also of great importance to architects. The revival of the classical style in architecture was largely based on one work, Vitruvius' *De architectura* ('On Architecture'), from the first century AD, and a renewed study of the classical buildings still standing. Early writers on perspective, such as Filipppo Brunelleschi (1377–1446) and Leon Battista Alberti (1404–1472), combined the practical mathematics of the masons and architects with geometrical constructions, but what is generally considered the first work on perspective specifically for the painter is *De prospectiva pingenti* ('On Perspective in Painting') by Piero della Francesca (*c.* 1412–1492).

Piero della Francesca was the son of a commodity dealer in Sansepolcro, near Florence, and, probably with a view to entering the family business, studied mathematics

▲ From the *Kalender of Shepherdes* (London, 1506), this woodcut is in stark contrast to contemporary pioneers of drawing in perspective.

◄ *The Measurers*, a Flemish painting of the sixteenth century showing a range of mathematical instruments. This has similarities with the Italian tradition of teaching practical mathematics in the so-called 'scuole d'abbaco'.

at one of the many schools of practical mathematics that were springing up in Italy at the time. He showed talent, and may well have become a mathematician by trade, but decided instead to become apprenticed to a local artist. His unique combination of skills makes Piero one of the few to sit comfortably in the annals of both art and mathematics. He seems to have spent very little time in Florence, and most of his famous works are to be found in fairly minor towns such as Urbino. Only three treatises have come down to us, with neither their precise date of composition nor their original titles known. Before looking at his work on perspective it is worth commenting on one innovation in geometry. Piero is credited with having rediscovered five of the Archimedean solids, which are so-called because in the fourth century AD the Alexandrian mathematician Pappus ascribed their discovery to Archimedes. Illustrated by Kepler in 1619, there are thirteen Archimedean solids in total, being an extension of the five Platonic solids to include faces constructed from more than one regular polygon. Five of the Archimedean solids are constructed by truncating the edges of the Platonic solids. Before Piero these solids were described rhetorically, merely by stating the required polygons, but Piero describes their construction and illustrates them. Not all the solids are illustrated in quite the right perspective, but this was an immense step forward at a time when works on practical geometry often illustrated solids schematically – for example, a cone would be illustrated as a triangle placed over a circle. Piero's work was to reappear in Luca Pacioli's *De divina proportione* ('On Divine Proportion'). Published in Venice in 1509, it included illustrations by Pacioli's friend Leonardo da Vinci (1452–1519) and a sixth Archimedean solid, the rhombicuboctahedron.

Piero's 'On Perspective' survives as fifteenth-century manuscripts in both Latin and the Tuscan vernacular. The introduction states that it is concerned only with the use of perspective for painters. But Piero and his contemporaries saw the rules of perspective as part of the broader science of optics. The constructions are not only about creating naturalistic pictures – the point is that, for them to appear natural, they must obey the rules that govern how the eye sees the world. The eye of the beholder is therefore central to the whole work. If a painting is a snapshot of a scene seen through a window, there is only one point in space from which the viewer has the correct view. The viewer's eyes must be at the same height as the horizon of the picture and focusing on the vanishing point. The transversals, which aid in constructing the foreshortening of objects

▲ Piero della Francesca, *The Flagellation of Christ*, which shows many features from Piero's treatise on perspective, including the chequered pavimento and architectural elements.

with distance, meet at a point on the horizon. This point is normally outside the frame of the image, and the distance between this point and the vanishing point is the optimal viewing distance away from the plane of the picture. 'On Perspective' is written in Euclidean fashion, laid out as theorems and proofs in the style of the *Elements*. Piero presents a number of constructions which map the real 'perfect' figure onto the plane of the picture, thereby creating the 'degraded' figure as it should be represented on that plane, with the lines of sight coming to a focus at the viewer's eye. He proceeds from the construction for a square floor to a tiled floor, a *pavimento*, to illustrate how the tiles farther from the viewer would foreshorten with increasing distance. He then considers other polygons, giving both their proper shape and their 'degraded' forms as seen at an angle. Piero proceeds to deal with prisms, from a cube to various column shapes such as a

➤ Piero della Francesca, *Annunciation, Madonna and Child with Saints*. We see here the rigorous use of perspective and the religious requirement that the figures are slightly larger than one would expect given the architectural structures.

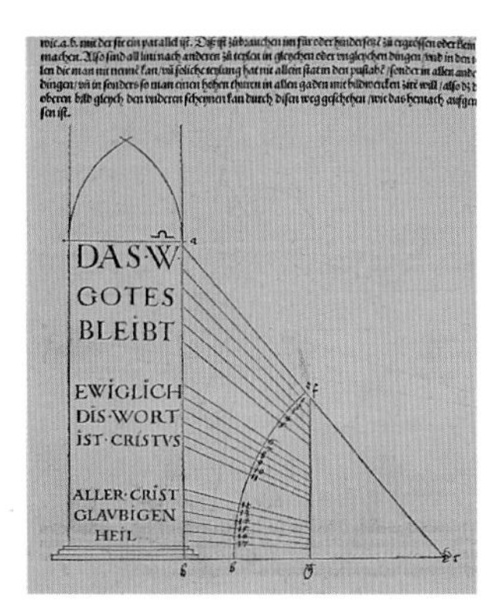

▲ Albrecht Dürer's *Treatise on Measurement with Compasses and Straightedge* (Nuremberg, 1525), showing the change in lettering size on a column such that they are legible from the ground.

➤ Michelangelo, *The Last Judgement*. At ground level, the larger upper figures would appear the same size as those below, using the same construction as the Dürer column on this page.

hexagonal prism, and on to how a long row of columns would look in perspective. He ends with a series of images of the human head from a number of different viewpoints.

Piero's work was later elaborated on and used by both painters and architects, as well as set designers for the theatre. The effect of perspective on the paintings of the era is somewhat subject to debate. We see it employed before Piero in pictures such as Domenico Veneziano's *The Annunciation* and Paolo Uccello's *Rout of San Romano*. We see it in Piero's *The Flagellation of Christ*, which can be seen as the practical embodiment of his treatise, but in his own *The Annunciation* we see that in the service of religious imagery figures are often much larger than they would be in a purely naturalistic painting in order to emphasize their importance. Michelangelo claims to have little time for mathematical precision, relying on the 'compasses in his eyes'. However, the Sistine Chapel is certainly painted in strict accordance with perspective; and in *The Last Judgement* Michelangelo renders the figures in the upper part of the picture much larger than those lower down to allow for the fact they would be viewed from a greater distance, an aim not apparent when seeing the fresco as an image in a book. Although artists very quickly learnt this new technique, artistic vision was not sacrificed to mathematical purity.

By the sixteenth century, Piero was remembered more as a mathematician than as an artist. His treatise was never published during the Renaissance, but it circulated in manuscript form and its content found its way into the publications of others. But to many his constructions of the more complicated figures were hard to follow, and these more demanding sections were often ignored. Interest did grow in building instruments, similar to those used by surveyors, to assist artists in representing objects in perspective. Albrecht Dürer's *Treatise on Measurement with Compasses and Straightedge* illustrates a number of such instruments. In most of them a stretched string representing the line of sight intersected a frame with movable cross-wires; the image was built up point by point. Alternatively, the artist could view the scene through a square grid, which gave an effect something like a coordinate system. Such a device was already used as a way of scaling up drawings to be painted.

Albrecht Dürer (1471–1528) was one of 18 children, born in Nuremburg to Hungarian parents, and apparently destined to follow his father into the jewellery trade. But by the age of thirteen it was clear that he was a very skilled artist and he later became an apprentice painter and woodcut designer. The early 1490s Dürer spent travelling and developing his idea of a new art based on the science of mathematics. On his return to Nuremberg he began studying the works of Euclid, Vitruvius, Pacioli and Alberti. He was later to visit Pacioli in Bologna, and planned a major work on mathematics and art of his own. By the time of his famous etching *Melancholia* (1514), his fame had spread. He had received high-profile commissions from Frederick the Wise, Elector of Saxony, and Maximillian I, the Holy Roman Emperor, and he owned a thriving printing company. He completed his *Treatise on Proportions* in 1523, but thought the mathematical contents

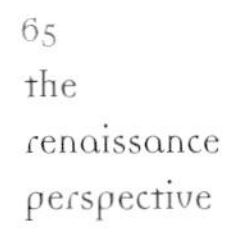

▲ A regular dodecahedron, the Platonic symbol of the cosmos, as illustrated by Leonardo da Vinci in Luca Pacioli's *De divina proportione* (1509).

too advanced for his readers, and so set about editing it down to the more accessible *Treatise on Measurement*, publishing it in 1525. Apart from earlier works on commercial arithmetic, this was the first mathematics book to be printed in German, making Dürer one of the most important Renaissance mathematicians. The work was largely on plane and solid geometry, including methods of construction and a section on perspective. An important part of the work is devoted to the depiction of solid objects in plan and elevation, a branch of mathematics now referred to as descriptive geometry, but also of very practical concern to architects and engineers.

The marriage of perspective geometry with the conic sections – those slices through a cone which produce figures such as a circle, ellipse, and parabola – led to another new branch of mathematics, projective geometry. Girard Desargues (1591–1661), a wealthy and learned Lyonnais, published little during his lifetime but kept up with mathematical developments through the circle of correspondents maintained by the mathematician, priest and philosopher Marin Mersenne. In 1639 he did publish a treatise known as *Rough Draft on Conics*. This was a difficult text to follow, and just 50 copies were produced for private distribution.

The basis of perspective geometry is that from the viewer's vantage point the 'perfect' and 'degraded' figures appear identical. Extending this result beyond the plane of a painting means that an original image can be projected onto an infinite number of planes and still appear unchanged to a fixed observer. Desargues investigated which properties of figures did remain unchanged, or invariant, under such projective transformations. One of his achievements was to unify the conics by treating them as projective transformations of the circle along the light cone – a tilted circle was indeed an ellipse.

The beauty of this approach is that, having established a theorem for one conic, say the circle, it was then simply a matter of effecting the appropriate projection and reinterpreting the theorem. But Desargues's achievement was in developing a new method rather than in establishing ground-breaking theorems. At the same time, Descartes's algebraic geometry was proving such a powerful tool that he suggested that Desargues's work would be more comprehensible if it were translated into his algebraic notation. Descartes later conceded that this was perhaps a quibble over style rather than content. But mathematicians were marching in an altogether different direction, and Desargues's work languished. His projective geometry and Dürer's descriptive geometrywere both to re-establish themselves on a sound mathematical footing later, in the early nineteenth century.

But when great and ingenious artists behold their so inept performances, not undeservedly do they ridicule the blindness of such men; since sane judgment abhors nothing so much as a picture perpetrated with no technical knowledge, although with plenty of care and diligence. Now the sole reason why painters of this sort are not aware of their own error is that they have not learnt geometry, without which no one can either be or become an absolute artist; but the blame for this should be laid upon their masters, who are themselves ignorant of this art.

Albrecht Dürer, *The Art of Measurement*, 1525

chapter ten

mathematics for the common wealth

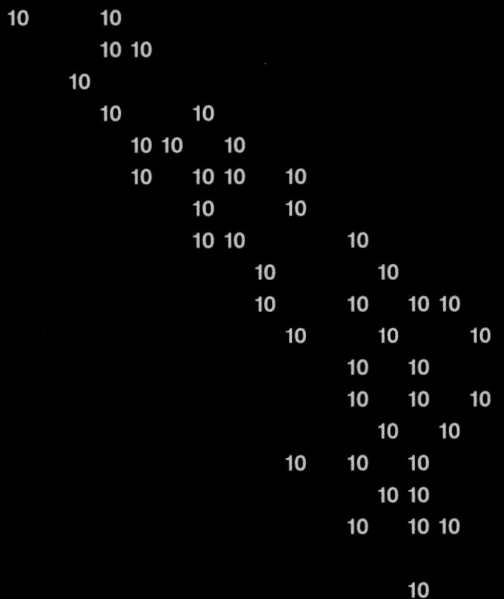

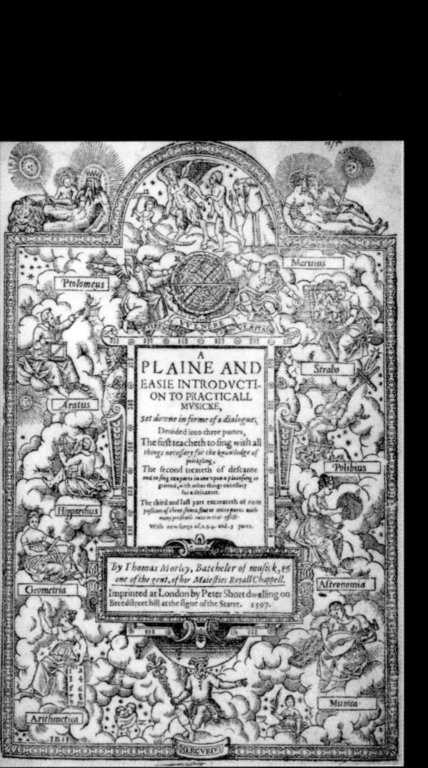

A PLAINE AND EASIE INTRODVCTION TO PRACTICALL MVSICKE,

Set downe in forme of a dialogue:

Deuided into three partes,

The first teacheth to sing with all *things necessary for the knowledge of pricktsong.*

The second treateth of descante *and to sing two parts in one vpon a plainsong or ground, with other things necessary for a descanter.*

The third and last part entreateth of composition *of three, foure, fiue or more parts with many profitable rules to that effect.*

With new songs of 2. 3. 4. and 5 parts.

By Thomas Morley, Batcheler of musick, & one of the gent. of her Maiesties Royall Chappell.

Imprinted at London by Peter Short dwelling on Breedstreet hill at the signe of the Starre. 1597.

◄ This late sixteenth-century frontispiece was widely used, most famously in Henry Billingsley's 1570 translation of Euclid's *Elements* into English with its *Mathematicall praeface* by John Dee. This illustration is from Thomas Morley's treatise on music.

Sixteenth-century Europe held the promise of endless possibilities. The previous two centuries had convulsed the continent with a series of natural and man-made disasters: the Black Death in the mid-fourteenth century wiped out virtually half the population, showing supreme indifference to social status and wealth; the Hundred Years' War between England and France left populations exhausted both physically and morally; and in 1453 the fall of Constantinople to the Ottoman Turks signalled the end of the Byzantine Empire. At the same time we see the efflorescence of the Italian Renaissance and the Humanist tradition, combining reverence for antiquity with a new-found confidence in personal freedom and education. The invention of printing and engraving meant that the new ideas could be disseminated more widely than could have been possible previously. Europe was also looking outwards to the rest of the world, and sea voyages of discovery, conquest and trade were on the increase. But navigation required accurate charts of the seas and the heavens, and trade required an efficient form of accountancy – none of these were sufficiently advanced at the time. Algebra, trigonometry, geometric projections, logarithms and calculus were all about to expand or make their first appearance. Before looking at these developments, it is worth considering the rising status of mathematics at the time.

As we saw earlier, mathematics was an integral part of scholastic training in the monasteries with the quadrivium of arithmetic, geometry, harmony and astronomy. But slavish reverence of ancient texts and the narrow mathematical requirements of the ecclesiastical authorities limited what could be achieved within this scholastic tradition. The term *mathematicus* was used to denote either a mathematician or an astrologer (Kepler complained that he earned far more from calculating astrological charts than he did from his astronomical work). Although there was as yet no such thing as a professional mathematician, economic growth in Europe created a need for a broad spectrum of numerate individuals who could handle financial and commercial affairs. These posts were to be filled from the guilds and workshops of craftsmen rather than from the universities. In the Renaissance, the sons of the mercantile classes would get a thorough grounding in elementary mathematics in schools or workshops. It was here that the use of Hindu-Arabic numerals became popular.

The new numbers had begun to be assimilated in the twelfth century with translations into Latin of Arabic texts, and in 1202 we find the publication of *Liber abbaci* ('Book of the Abbacus') by Leonardo of Pisa (*c.* 1180–*c.* 1250), known also as Fibonacci. Now regarded as a mathematical landmark, it was not as popular in its time as the more elementary *Algorismus vulgaris* by John of Halifax (*c.* 1200–1256), better known as Sacrobosco. The title *Liber abbaci* is unfortunately rather misleading. The term 'abbacus', with two 'b's, refers to the calculating methods using the new numerals, and has nothing to do with the calculating device known as the abacus. Indeed, there was rivalry between proponents of the two forms of computation, and it is better to use the term 'algorist' for a practitioner of

➤ Robert Recorde's *The Castle of Knowledge* (1556) is his text on cosmology. This frontispiece illustrates Recorde's pedagogical aims and the triumph of reason over authority, with Ignorance standing precariously upon a sphere in contrast to Knowledge standing squarely and assuredly.

▲ French sixteenth-century text illustrating the theory and use of the cross staff, an instrument used for measuring the altitude of the Sun and Pole Star so that navigators could calculate their latitude at sea.

abbacus techniques, and 'abacist' for someone who still preferred the abacus or counting board. One proficient in the rules of the abbacus was known as a *maestro d'abbaco*.

In *Liber abbaci*, Fibonacci devoted considerable space to mercantile mathematics. In international trade, merchants had to cope with dozens of different systems of weights and measures, and handle transactions in different currencies, and they needed an efficient method of calculation to avoid serious errors. In 1494 Luca Pacioli published his *Summa*, now famous as the first work on accountancy practices such as double-entry bookkeeping, it was also a compendium of useful mathematics of the period, including arithmetic, algebra and geometry. The earliest printed arithmetic had been published a few years before in 1478, anonymously in Treviso. Notation was still fluid during this period, with fractions still expressed in sexagesimal notation or unit fractions. Decimal fractions became popular in the sixteenth century, though sexagesimals persisted in astronomical calculations, and the decimal point was made popular by Napier.

The tendency to write in the local vernacular in preference to Latin made mathematical textbooks accessible to a wider public, although it simultaneously hindered their dissemination across language barriers. Adam Riese was influential in promoting Hindu-Arabic numerals in the German language. Robert Recorde (*c.* 1510–1558) was probably the first popularizer of mathematics. He wrote the first mathematical textbooks in his native English, and his work on arithmetic, *The Ground of Artes* (1543), remained in print for over a hundred and fifty years. Most of his books were written in dialogue form, and included diagrams and examples to further his pedagogical aims – in many ways he was a one-man distance learning course. His most quoted work is *The Whetstone of Witte* (1557), an elementary algebra text in which we find the first use of =, the equals sign.

Sith Merchauntes by shippes great riches do winne,
I may with good righte at their seate beginne.
The shippes on the sea with saile and with ore,
were first founde, and styll made, by Geometries lore.
Their compas, their carde, their pulleis, their Ankers,
were founde by the skill of witty Geometers.
To sette forth the capstocke, and eche other part,
wold make a great showe of Geometries arte.
Carpenters, carvers, Joiners and Masons,
painters and Limners with suche occupations,
Broderers, Goldesmithes, if they be cunning,
Must yelde to Geometrye thankes for their learning.
The carte and the plowe, who doth them well marke,
Are made by good Geometrye. And so in the warke
of Tailers and shoomakers, in all shapes and fashion,
The woorke is not praised, if it wante proportion.
So weavers by Geometrye trade their foundacion,
Their Loome is a frame of straunge imaginacion.
The wheele that doth spinne, the stone that doth grind,
The Myll that is driven by water or wince,
Are workes of Geometrye straunge in their trade,
Fewe could them devise, if they were unmade.
And all that is wrought by weight or by measure,
without proofe of Geometry can never be sure.
Clockes that be made the times to device,
The wittiest invencion that ever was spied,
Now that they are common they are not regarded,
The artes man contemned, the woorke unrewarded.
But if they were scarse, and one for a shewe,
Made by Geometrye, then should men know,
That never was arte so wonderfull witty,
so needfull to man, as is good Geometry.

Robert Recorde, *The Pathway to Knowledge,* 1551

In these words we can detect the two contrasting views of mathematics that have held throughout history: mathematics as a utilitarian tool, and as an aesthetic study. Recorde was also a staunch advocate of the right of reason over authority, regarding mathematics as the noble art for seeking true knowledge. Such an attitude may not have been universally popular, and although he held the post of Comptroller of the Bristol Mint and Surveyor of the Mines and Monies of Ireland, Recorde ended his days in prison, most probably for political indiscretions.

A contemporary and colleague of Recorde's, John Dee (1527–1608), had a similarly high-flying career followed by a vertiginous fall from grace. They were both consultants on navigation and cartography to the Muscovy Company, and Dee wrote *The Perfect Arte of Navigation* in 1577. But his main interests lay in the occult sciences, being the leading Elizabethan light in the Renaissance Neoplatonist tradition, and his studies included the Kabbalah (the Jewish mystical tradition) and alchemy. He held the position of Royal Astrologer to Queen Elizabeth, casting horoscopes and advising on calendar reforms. But his reputation made him both admired and feared at court, and although he had been an adviser to Elizabeth from before her coronation, he sensed that her patronage was often at arms length. He often felt the need to defend himself publicly, pleading that his studies were very much for the benefit of the realm. Indeed, on returning from travels in Europe he was promised a pension, but it never materialized and he died destitute in 1608. One of the most famous proclamations on the value of mathe-matics is to be found in Dee's preface to Euclid's *Elements* as translated by Henry Billingsley, later to became Lord Mayor of London. This is the first scholarly edition of the *Elements* in English, and was probably edited by Dee himself.

John Napier was not a professional mathematician but a Scottish laird, Baron of Murchiston, and spent much of his time managing his estate. But he found time to write on a variety of topics, and got himself embroiled in anti-popish theology. Although the Hindu-Arabic numerals were in common use by then, calculations were performed with pen and paper, and people sought ways of speeding up sometimes lengthy procedures. Napier is credited with two inventions which greatly facilitated calculations – Napier's bones, and logarithms. Napier's bones, also known as Napier's rods, were sticks on which were carved multiplication tables and could be arranged in a lattice pattern so that one could quickly read off any lengthy multiplication. The rods essentially turned lengthy multiplications into simple additions. The invention of logarithms was similarly inspired by a desire for greater speed. The word itself was coined as a compound of *logos* ('ratio') and *arithmos* ('number'). Many mathematicians had been struck by the relationship between arithmetic and geometric series, and that the product of two powers could be reduced to the sum of the powers. Napier's insight was that this could apply to any power, and he compiled a table of Napierian logarithms which appeared in his 1614 book *Mirifici logarithmorum canonis descriptio* ('A Description of the Marvelous Rule of Logarithms').

▲ Representation of Arithmetic from Gregor Reisch's *Margarita Philosophica* (1503). The transition from the Roman abacus to the use of Hindu-Arabic numerals was surprisingly slow, with centuries of rivalry between the two systems. Indeed the new numerals were so novel that, if you turn the image upside-down, you'll note that the artist has illustrated a meaningless calculation!

His original idea did not really make use of a number base: instead, he divided the unit number line into 10^7, parts which gave enough significant figures for most calculations. He then defined the relationship $N=10^7(0.9999999)L$, where L is the logarithm of N. This gave the logarithm of 10^7 as 0, and that of 9999999 as 1; numbers in between took values between 0 and 1. His tables gave the logarithms of trigonometric functions rather than natural numbers, illustrating his primary concern with the tedious calculations necessary in astronomy and navigation. One of Napier's great admirers was Henry Briggs, first Savilian Professor of Geometry at Oxford. They both agreed that a more practical table could be constructed by setting $\log 1=0$ and $\log 10=1$. But Napier died in 1617, and it fell to Briggs to compile the first logarithmic table to base 10 which would be recognizable today. This table was for the numbers 1 to 1000; in 1624 Briggs extended it to 100,000, both sets of logarithms accurate to fourteen places. The advantage of having a fixed base was that the removal of the factor 10^7 from calculations revealed the fundamental rule of logarithms – that the logarithm of a product of two numbers is equal to the sum of the individual logarithms. Today's calculators have made tables of logarithms, trigonometric functions and reciprocals redundant, along with slide rules, but at the time Briggs's tables were heralded as a great labour-saving device. Navigators, who had to handle sines and cosines, found that a typical task of multiplying two seven-figure numbers was reduced to consulting their logarithms, making one addition, and then referring back to the table where the inverse logarithm would give the required answer. Before, when a calculation could take you an hour, your answer would always lag an hour behind your real position. This had now been cut down to a few minutes.

Francis Bacon (1561–1626) was neither a mathematician nor a scientist and yet, rather like Plato, he had enormous influence on the philosophy of the scientific enterprise. During Queen Elizabeth's reign he served in the Commons and was a Queen's Counsel, although without a true portfolio. It was on the accession of King James I that his career took off, with a succession of influential positions culminating in his appointment as Lord

◄ Hans Holbein the Younger, *The Ambassadors* (1533). The French ambassadors were at the court of Henry VIII to dissuade him from divorcing Catherine of Aragon. The range of mathematical instruments symbolize learning as enshrined in the quadrivium as well as the power such knowledge confers.

▼ A page from the 1628 edition of John Napier's *Arithmetica logarithmica*, completed and published posthumously by Henry Briggs, the first Savilian professor of geometry at Oxford. This edition was later used by Charles Babbage in his analysis of tabular errors.

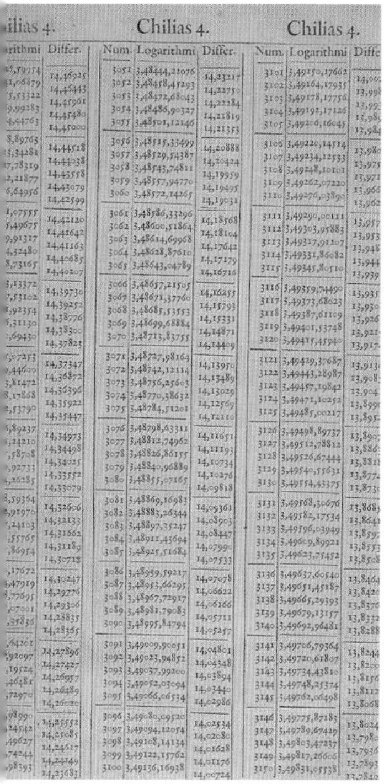

Chancellor in 1618. In an age when patronage was endemic it seems strange that Bacon should have been impeached for bribery in 1621. Despite this, James continued to pay his pension, and the fall seems to have hurt his pride more than his pocket. His publications spearheaded the drive to make natural philosophy a primary issue for both the government and the Crown. *The Advancement of Learning* (1605) and *The Great Instauration* (1620) were dedicated to James, as a call to him to become a patron of science. Bacon's writings influenced later scientists such as Newton and Halley, and are credited with providing the English cornerstone for the scientific revolution, and with inspiring the founding of the Royal Society. His position also meant that science had a champion with political and financial influence. Knowledge was power, and science became valued as the engine for greater prosperity in the Common Wealth, a view expanded by Bacon in his *Novum organum* (1620). Bacon's views on mathematics were very much skewed towards the utilitarian – mathematics was the language of science and the tool at its disposal. But he also had the modesty and foresight to predict that mathematics itself was not a static discipline, and that new branches of study were almost certain to develop. The use of mathematics by merchants, navigators and scientists was seen as contributing to the creation of greater wealth for the nation. The promotion of mathematics was no longer the concern of a few scholars, but a full-blown call to arms.

No state, no age, no man, nor child, but here may wisdome win
For numbers teach the parts of speech, wher children first begin.
And number bears so great a sway even from the most to least
That who in numbring hath no skill, is numbred for a beast:
For what more beastly can be thoght? nay what more blockish than
Then man to want the onely art, which proper is to man,
For many creatures farre excell mankind in many things,
But never none could number yet, save man in whom it springs.
If numbring then be (almost) al, betweene a man and beast,
Come learne o men to number then, which arte is here profest
If martial man thou minde to be, or office do expect.
In court or country where thou dwelst, or if thou do elect,
In phisicke and philosophie, or law to spend thy dayes,
Assure thyself without this arte, thou never canst have praise.
I overpasse Astronomie, and Geometrie also,
Cosmographie, Geographie, and many others mo,
And musick with her dulcet tunes, all which without this arte,
Thou never canst attayne unto, nor scarce to any part,
Ne canst thou be an auditor, or make a true survey,
Nor make a common reckoning, if numbers be away.

But if thou willte a merchant be, then make this booke thy Muse,
wher thou shalt find rules fit for thee, as thou canst wish or chuse.
And having onely handie craft, yet herein mayst thou finde,
such things as may oft serve thy turne, and much enrich thy minde.
Nay if thou but a shepherd be, it wil thee sore accumber
To do thy duty as thou shouldst without the help of number.
To number all the benefits, that number brings to man,
would be too long here to rehearse, and more than well I can,
wherefore to speake one woorde for all, and let the rest alone,
without this art man is no man, but like a block or stone.

Thomas Hylles, *The Arte of Vulgar Arithmeticke,* 1600

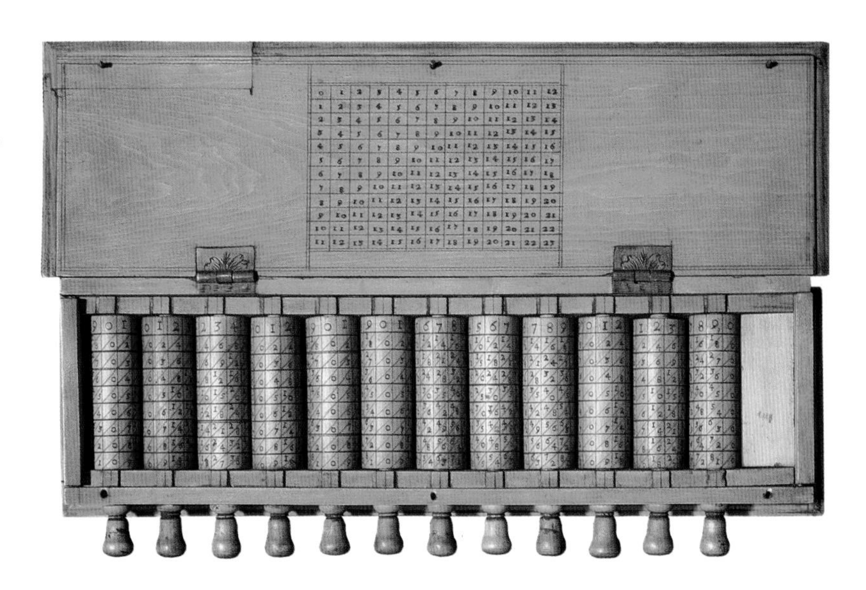

➤ John Napier's very popular aid to calculation, Napier's rods or 'bones' were initially made as four-sided ivory or wooden rods. In this later form the rods are mounted in a case and can be rotated. Essentially, the device turned a lengthy multiplication into a sequence of simple additions.

chapter eleven

the marriage of algebra and geometry

◄ Detail from Newton's *Opticks* (1704). A short treatise appended to the *Opticks* was the *Enumeration of Curves of Third Degree* in which Newton lists the seventy-two different species of cubics and draws curves for most of them. For the first time, we see two axes fixed at right angles and the use of negative coordinates.

From the Greek era onwards, mathematics was split into two main branches, geometry and arithmetic, one dealing with magnitudes, the other with numbers. But there could never be a truly clean break between the two, and we have seen how cultures gave relative prominence to one branch or the other depending on their particular concerns. The development of algebra and its relationship with geometry will be illustrated through the story of the solution of the cubic equation – what in modern notation is written as $ax^3 + bx^2 + cx + d = 0$.

The word al-jabr ('restoration'), taken from the title of al-Khwarizmi's treatise on algebra *Hisab al-jabr wa'l-muqabala* (Chapter 7), is the source of our word 'algebra'.

Al-Khwarizmi wrote his algebra in words, and stated the solutions to equations rhetorically. The powers of unknown quantities were given names, such as *shay* ('thing') for x, *mal* ('wealth') for x^2 and *ka'b* ('cube') for x^3. The names used for powers remained rather fluid, and in his *Liber abbaci* of 1202 (Chapter 10) Fibonacci used translations from the Arabic as well as some of his own, for example we find radix for root and cubus for x^3. The *Liber abbaci* was an important work in the transmission of the Hindu-Arabic numerals, describing the 'nine Indian figures' and the 'zephirum', or zero.

Al-Khwarizmi's text, written in the first half of the ninth century, divides solutions to quadratic equations into six types, restricting both the numerical coefficients and the final solutions to positive values (Chapter 7). Solutions are justified by geometric demonstrations which are essentially the same as the Babylonians' completing the square (Chapter 1). In the eleventh century, Umar ibn Ibrahim al-Khayyami (d. 1022) – known familiarly as Omar Khayyam – discovered a method of solving cubics geometrically as the points of intersection of two conic sections; for example, the solution to the equation $x^3 + ax = c$ can be found from the intersection of a circle and a parabola. Yet again, the coefficients and the solutions are all positive numbers. He did not find a general algebraic solution to the cubic, but his application of Greek geometry to the solution of algebraic equations was certainly sophisticated. In his own words, 'Algebras are geometric facts which are proved', and he hoped that a purely algebraic solution to the general cubic would be found by mathematicians coming after him. Unfortunately, al-Khayyami's *Algebra* does not seem to be one of the many Arabic books that were translated into Latin.

A general algebraic solution to the cubic – that is, a finite sequence of algebraic steps leading to complete solutions – was indeed found, but not until the Italian Renaissance, nearly four hundred years later. Approximate solutions were already known. In 1225, for example, Fibonacci published a treatise on the cubic which gave an approximate solution to a particular case, but unfortunately without any method. In looking at the story of the cubic we become immersed in the competitive world of Renaissance Italy. It was rare to publish new results, as withholding such discoveries increased one's reputation in the eyes of patrons. Exchanges took the form of competitions in which mathematicians

DE POLYGRAPHIE. 176

Ordredes antiques lettres Numerales.

DC	60000000	L	500000000
DCC	70000000	LX	600000000
DCCC	80000000	LXX	700000000
DCCCC	90000000	LXXX	800000000
X	100000000	LXXXX	900000000
XX	200000000	C	1000000000
XXX	300000000	C	20000000000
XXXX	400000000		

I'ay iuſques icy deduict, &deſcrit par forme d'exemple,le moyen & methode d'eſcrire en lettres Latines & communes, les nombres & la mode numerale des Anciens: à fin que d'iceux les ignorants en euſſent plus facile intelligence. Et quant à ce, qu'en la ſuſcrite deſcription & appoſition des figures d'algarithme,i'aurois en quelques endroicts laiſſé l'ordre& reigle de la vraye ſupputation & forme de cõpter,ce n'a eſté par erreur,ny auſsi par faulte ou omiſsion. Mais par bonne & raiſonnable cauſe i'ay aduiſé ainſi le faire:& meſmement à fin que par trop grand progreſsion & prolixité ne donnaſſe ennuy à moy,ny aux lecteurs,qui en la figure ſubſequente,briefuement cy apres tranſcrite,verront autres moyens & ordres numeraux, par leſquels les anciens manifeſtoient leurs grands nombres.

▲ This sixteenth-century text still feels it necessary to acquaint the reader with the relationship between Roman and Ordinary numerals. Indeed we still use Roman numerals today in a limited way.

would challenge one another with lists of questions, and winning such contests would further increase one's reputation.

The solution to the cubic, and indeed the quartic, was first published by Girolamo Cardano (1501–76) in his *Ars magna* (1545). But neither solution was the discovery of Cardano himself. The first real solution goes back to Scipione del Ferro (*c.* 1465–1526), professor of mathematics at Bologna. He never published his solution, but rather bequeathed it to a student of his, Antonio Maria Fior. Fior saw this as his ticket to fame and fortune, and challenged mathematicians to problem-solving contests. However, Fior seems to have been a fairly mediocre mathematician who relied on one solitary weapon. The mathematician Niccolò Fontana (*c.* 1500–57), better known as Tartaglia, the 'stammerer', by virtue of the speech defect inflicted by a sabre cut he received as a child in the French attack on the city of Brescia, was also working on the cubic. In 1535 Fior and Tartaglia met to do battle, and on the night of 12 February Tartaglia claimed to have also solved the cubic. Tartaglia won the competition, completing all Fior's problems, whereas Fior could not solve any of Tartaglia's.

At the time, the cubic was not thought of as a single equation, but was divided into types according to which terms were being equated, rather like al-Khwarizmi's quadratics. It therefore appears that Tartaglia had solved not only Fior's lone type, but also a number of other cubics. News of Tartaglia's victory reached the ears of Cardano, who eventually persuaded Tartaglia to divulge his secret in return for a letter of recommendation to a prospective patron. However, at their meeting in Milan in 1539 Tartaglia extracted an oath of secrecy from Cardano never to publish the solution, which was given to him in the form of a cryptic poem. But Cardano later discovered that del Ferro's son-in-law had in his possession the original manuscript and was granted permission to read it. He and his assistant Ludovico Ferrari (1522–65) had also made further strides in solving a general cubic as well as the quartic. Cardano gave due credit to Tartaglia's work, but having established del Ferro's priority he no longer felt bound by the pledge of secrecy. Tartaglia was incensed at this betrayal, and proceeded to seek revenge in a book stating his side of the story and in a long and bitter exchange in which Ferrari took up his master's cause, Ferrari himself being no mean mathematician. In 1548 Tartaglia was suddenly elevated from his lowly position as a mathematics teacher in Venice to a lectureship in Brescia. He decided that a challenge to Ferrari might bring further glory and some sweet revenge, but he had seriously underestimated Cardano's assistant, and hastily departed before the competition reached its conclusion. Things had seriously backfired on Tartaglia, and the authorities in Brescia subsequently refused to pay his stipend. He returned to Venice to resume teaching.

In contrast to Tartaglia, who was from a poor background and was always searching to secure patrons, Cardano was to achieve fame and a rather fragile wealth. Cardano was very much a man of his age – mathematician, physician, astrologer, gambler and heretic. He was refused entry into the College of Physicians for nearly fifteen years, allegedly because of his illegitimate birth but more likely because of his reputation for being outspoken and difficult. He nearly went bankrupt through his obsessive gambling, but he did manage to build up a flourishing private medical practice, and eventually lectured in medicine in Milan and Pavia from 1543 to 1552. He was summoned to Scotland to treat the Archbishop of St Andrew's. On his return he was made professor of medicine at Pavia University, his fame sealed with the news of the Archbishop's recovery, but his worldly success was to be undermined by his tempestuous family life. He could not save his favourite elder son from execution on a charge of poisoning his money-grabbing wife, her family demanding an extortionate amount in compensation from Cardano. He had to leave Pavia, and was made professor at Bologna. Then his younger son stole from his father's house to pay yet another gambling debt. This time an exasperated Cardano reported his son, who was banished. Cardano had not made many friends in Bologna, and in 1570 he was arrested for heresy, having cast the horoscope of Jesus Christ and praised the emperor Nero. Perhaps surprisingly, he was subsequently welcomed in Rome, and the Pope agreed to grant him a pension. His gambling habit had mixed fortunes, but it did provide the basis for a book on probability. His autobiography is a frank testament to an amazing life on the cusp of a mathematical revolution.

Cardano's successful attack on the cubic amounted essentially to a geometric 'completing the cube', analogous to the method of completing the square. However, the exposition was still in the style of al-Khwarizmi, with lengthy rhetorical explanations and a reliance on different classifications of the cubic as negative coefficients were still not regarded as legitimate. By transforming the more complicated cubics into simpler solvable types, he was thus able to go beyond del Ferro and Tartaglia. Cardano also noticed that sometimes the intermediate steps in the solution required him to take the square root of a negative number. He displayed a certain intellectual squeamishness when confronted with these complex numbers, but though he thought such answers meaningless he did not dismiss them entirely out of hand. On one occasion he obviously held his nerve long enough to find that on multiplying what we now call complex conjugates he obtained a real answer. He gave the conditions for which a cubic has complex solutions, but did not investigate these new types of number any further. In 1572, Rafael Bombelli (1526–*c.* 72) published his *Algebra*, in which he extended the field of numbers to square roots, cube roots and complex numbers. He also took a major step in the algebraic solution of geometric problems, and vice versa, but unfortunately it had little effect on his contemporaries as an important part of the work was left out by his editors, and published only in the twentieth century.

In Europe, the development of algebra went hand in hand with the use of the new Hindu-Arabic numerals. In 1494 the friar Luca Pacioli published his *Summa de arithmetica, geometrica, proportioni et proportionalita*, considered to be the first book on algebra. Pacioli's algebra is still largely a mixture of rhetorical and algebraic explanations (what is called 'syncopated'). The unknown in the equation was often referred to as *cosa* in Latin, and then *coss* in German. The so-called cossic art developed rapidly in early sixteenth-century Germany through works such as *Die Coss*, written in 1524 by the celebrated *Rechenmeister* Adam Riese (1492–1559). Many of

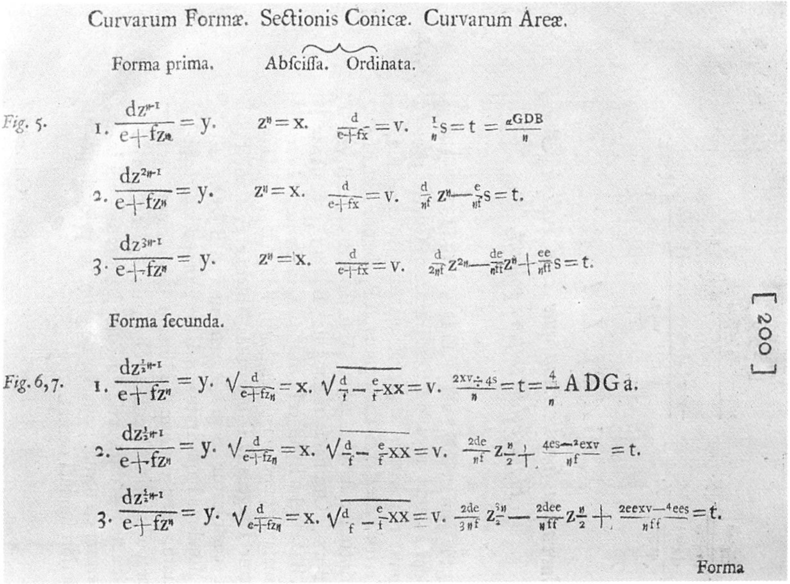

Curvarum Formæ. Sectionis Conicæ. Curvarum Areæ.

Forma prima. Abſciſſa. Ordinata.

Fig. 5.

1. $\frac{dz^{\eta-1}}{e+fz^{\eta}} = y.$ $z^{\eta} = x.$ $\frac{d}{e+fx} = v.$ $\frac{1}{\eta}s = t = \frac{\alpha GDB}{\eta}$

2. $\frac{dz^{2\eta-1}}{e+fz^{\eta}} = y.$ $z^{\eta} = x.$ $\frac{d}{e+fx} = v.$ $\frac{d}{\eta f}z^{\eta} - \frac{e}{\eta f}s = t.$

3. $\frac{dz^{3\eta-1}}{e+fz^{\eta}} = y.$ $z^{\eta} = x.$ $\frac{d}{e+fx} = v.$ $\frac{d}{2\eta f}z^{2\eta} - \frac{de}{\eta ff}z^{\eta} + \frac{ee}{\eta ff}s = t.$

Forma ſecunda.

Fig. 6, 7.

1. $\frac{dz^{\frac{1}{2}\eta-1}}{e+fz^{\eta}} = y.$ $\sqrt{\frac{d}{e+fz^{\eta}}} = x.$ $\sqrt{\frac{d}{f} - \frac{e}{f}xx} = v.$ $\frac{2xv \div 4s}{\eta} = t = \frac{4}{\eta}ADGa.$

2. $\frac{dz^{\frac{1}{2}\eta-1}}{e+fz^{\eta}} = y.$ $\sqrt{\frac{d}{e+fz^{\eta}}} = x.$ $\sqrt{\frac{d}{f} - \frac{e}{f}xx} = v.$ $\frac{2de}{\eta f}z^{\frac{\eta}{2}} + \frac{4es - 2exv}{\eta f} = t.$

3. $\frac{dz^{\frac{1}{2}\eta-1}}{e+fz^{\eta}} = y.$ $\sqrt{\frac{d}{e+fz^{\eta}}} = x.$ $\sqrt{\frac{d}{f} - \frac{e}{f}xx} = v.$ $\frac{2de}{3\eta f}z^{\frac{3\eta}{2}} - \frac{2dee}{\eta ff}z^{\frac{\eta}{2}} + \frac{2eexv - 4ees}{\eta ff} = t.$

Forma

[200]

➤ Newton's *Enumeration of Curves of Third Degree* was a triumph of both algebraic geometry and analytical geometry, using the calculus to discover new properties of the curves. Here we see algebraic expressions for the areas bounded by the curves under investigation.

the symbols we now recognize as algebraic start to be used in this period, the + and − signs coming from Germany, the = sign from England. The whole transition, from a rhetorical algebra through various individual syncopations to a standardized and unambiguous symbolic algebra, took a few hundred years. One major preoccupation was the role of powers higher than three. As algebraic methods relied on geometric proofs, and there were no physical dimensions higher than three, it did not seem reasonable to ascribe any meaning to a fourth power or higher. The very terms that were used highlight this problem, the fourth power usually being referred to as 'square-square'. In the mid-sixteenth century Robert Recorde felt obliged to give a lengthy justification for using higher powers, appealing to the fact that the area of a square whose sides are themselves square numbers is effectively a number raised to the fourth power, hence the term square-square.

The break from a purely geometrical approach came with the publication of *La Géométrie* by René Descartes (1596–1650). This important work was a mere appendix to the *Discours de la méthode pour bien conduire sa raison et chercher la verité dans les sciences*, and was often dropped from subsequent editions. Descartes's aim in the *Discours* was to set out a philosophy of science that would lead to correct knowledge about a universe of matter and motion. A correct description of the universe in the language of mathematics therefore required that the language itself be based upon solid foundations. Despite its name, *La Géométrie* is essentially a marriage of algebra and geometry, what is now known as analytic geometry; it proves the equivalences between geometric constructions and algebraic manipulations, and curves are described by equations. Descartes also broke with tradition in regarding powers as numbers rather than geometric objects: x^2 was no longer an area but a number raised to the second power; its geometric equivalent was the parabola, not the square. This liberated algebra from the obligation of dimensional homogeneity, a restriction which required every term in an equation to have the same dimension. We find, for example, expressions such as *xxx* + *aax* = *bbb*, so that each term represents a cubic quantity. Indeed, Descartes was happy to discuss curves of any power x^n. This shift was so powerful that in mathematics we no longer even think of x^2 as an actual square. His algebra looks familiar to the modern eye, using letters at the start of the alphabet for coefficients and those at the end of the alphabet for variables, the only symbol that appears incongruous is the use of ∞ for =.

> If, then, we wish to solve any problem, we first suppose the solution already effected, and give names to all the lines that seem needful for its construction — to those that are unknown as well as to those that are known. Then, making no distinction between known and unknown lines, we must unravel the difficulty in any way that shows most naturally the relations between these lines, until we find it possible to express a single quantity in two ways. This will constitute an equation (in a single unknown), since the terms of the one of these two expressions are together equal to the terms of the other.
>
> Descartes, *La Géométrie*, 1637

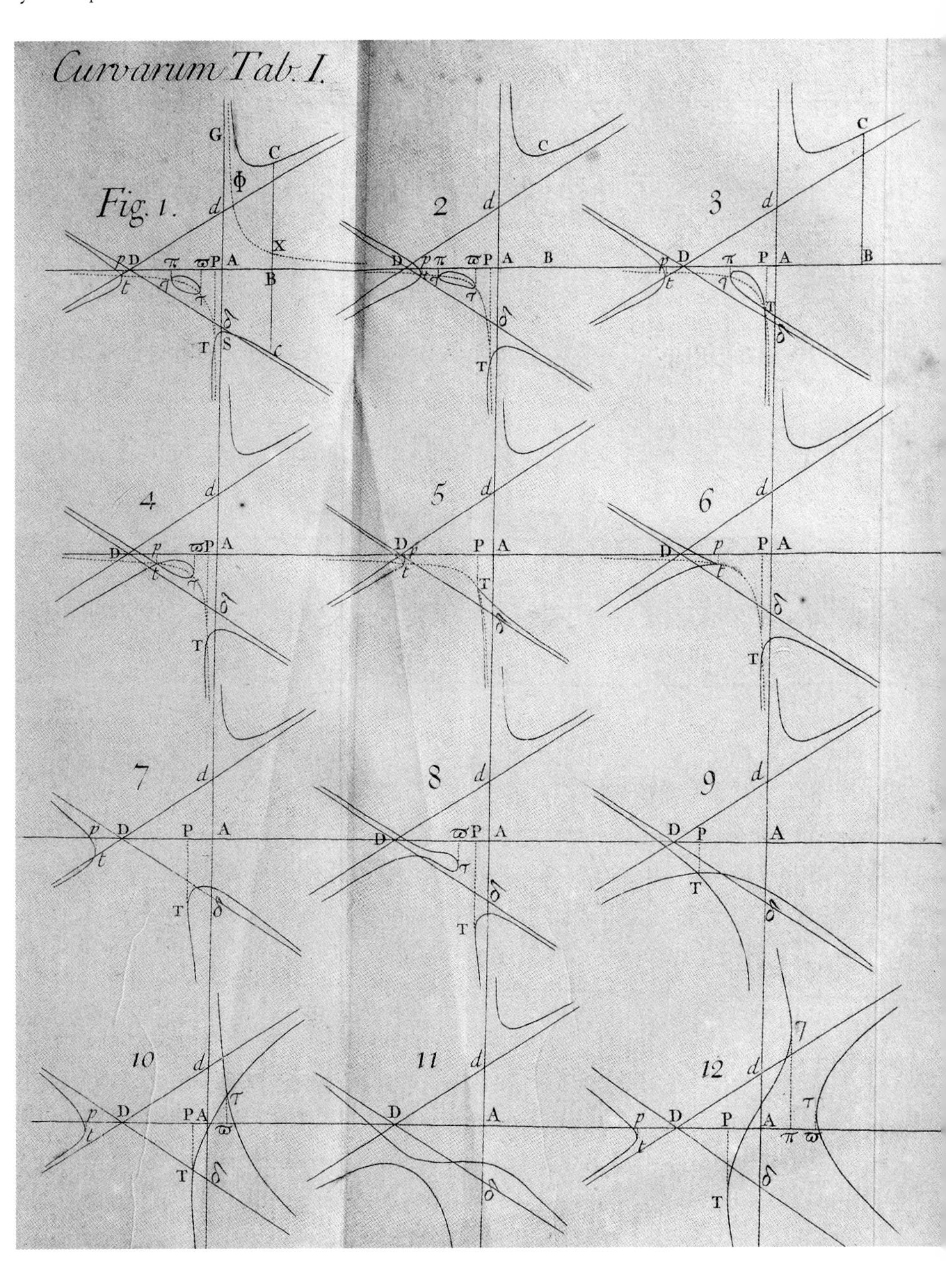

➤ The *Enumeration of Curves of Third Degree*, an appendix to Newton's *Opticks* (1704), shows that the marriage of algebra and geometry had reached a form that is recognizably modern. Each point on a curve was denoted by a coordinate (x,y) whose value satisfied the particular equation being investigated.

The cubic could now be solved as intersecting conics, analogous to al-Khayyami's method, but now one could actually construct a cubic equation. Descartes took pains to relate algebraic manipulations to geometric transformations, so that Cardano's formula was no longer a matter of 'completing the cube' but of transforming the cubic curve. Descartes thus also freed geometry from the restrictions of using ruler and compass constructions. There are many aspects of what is now known as algebraic geometry which are not found in Descartes's *La Géométrie*, such as coordinate axes, formulas for distances between points or angles between lines. Given the title of the work within which *La Géométrie* is found, it is perhaps fitting that Descartes's importance is in giving future mathematicians a new method or language in which to phrase mathematical problems, and a certain parity between algebraic and geometric methods.

When the cube and the things together
Are equal to some discrete number,
Find two other numbers differing in this one.
Then you will keep this as a habit
That their product should always be equal
Exactly to the cube of a third of the things.
The remainder then as a general rule
of their cube roots subtracted
will be equal to your principal thing.
In the second of these acts,
when the cube remains alone,
You will observe these other agreements:
You will at once divide the number into two parts
so that the one times the other produces clearly
The cube of a third of the things exactly.
Then of these two parts, as a habitual rule,
You will take the cube roots added together,
And this sum will be your thought.
The third of these calculations of ours
Is solved with the second if you take good care,
As in their nature they are almost matched.
These things I found, and not with sluggish steps,
In the year one thousand five hundred, four and thirty.
With foundations strong and sturdy
In the city girdled by the sea.

Solution to the cubic as given to Cardano by Tartaglia in 1546

chapter twelve

the clockwork universe

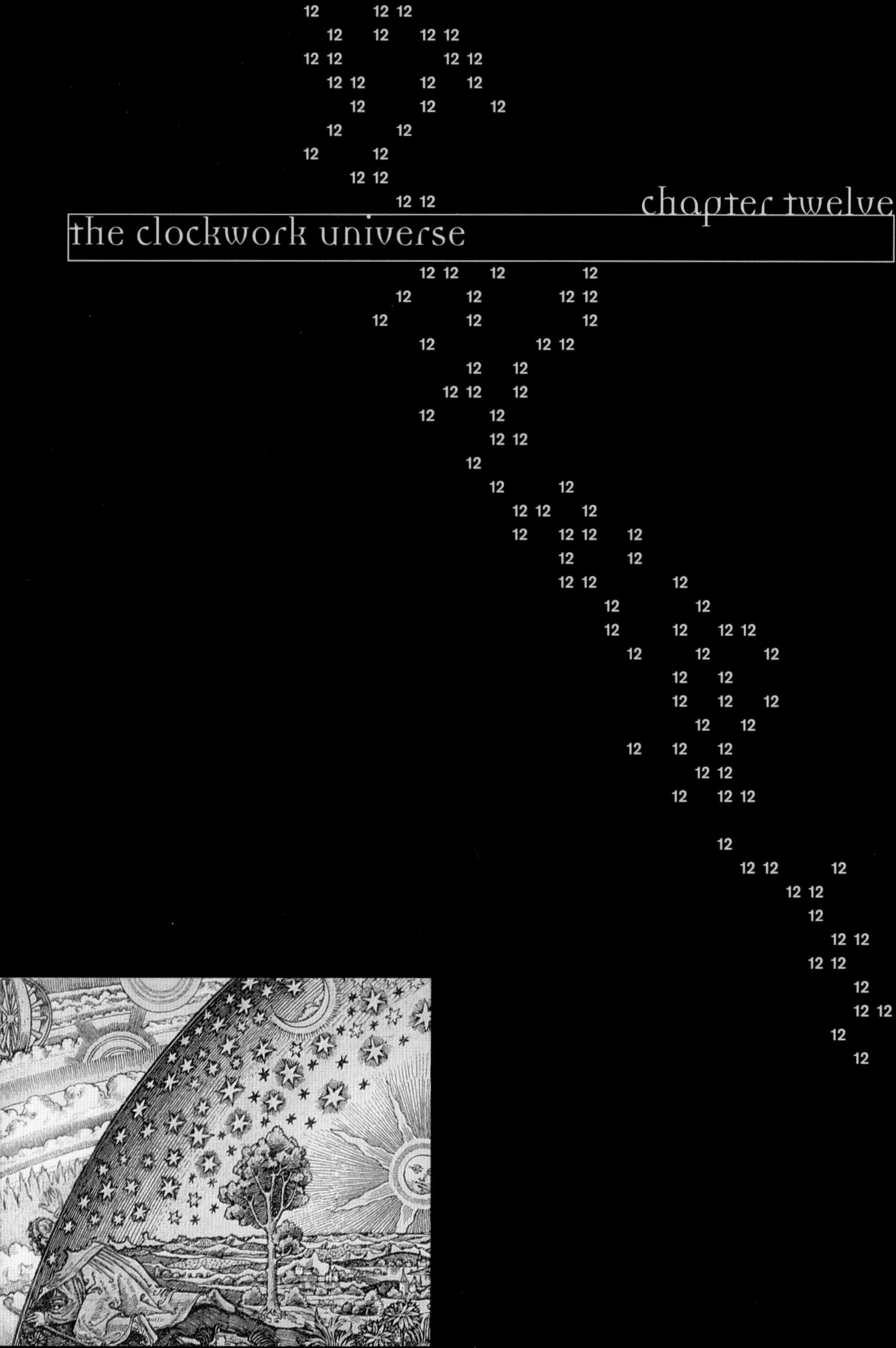

◀ From Camille Flammarion, *L'atmosphere meteorologie populaire* (Paris, 1888). This woodcut in the style of the early sixteenth century represents breaking through the medieval world to see the underlying mechanisms which turn the universe.

In the sixteenth century, Ptolemy's *Almagest* (see Chapter 2) remained the primary source of information on planetary orbits. The Ptolemaic system's cumbersome framework of epicycles and deferents survived in various forms for nearly two thousand years, probably because both the trigonometric tables used and the observational data collected were not accurate enough to highlight the deep flaws in the system. Aristotle's glass spheres in perfect circular motion were replaced by a host of angels – heavenly spirits turning the heavenly bodies. For Ptolemy the mathematics was there to 'save the phenomenon', not to explain it, and he successfully combined Aristotle's philosophical demands with the observable facts. The revolution about to take place would, literally, move Heaven and Earth. A crucial aspect of this revolution was the role of mathematics – does an accurate mathematical model tell us something about physical reality?

One of the most obvious problems with the Ptolemaic system is that, as a planet moves around its epicycle, its distance from Earth varies considerably, so its apparent size in the sky should change. Such change is most obvious in the case of the Moon, and it was probably this concern that prompted Nicolas Copernicus (1473–1543) to propose

▼ Cellarius, *Atlas Coelestis*, 1660, illustrating the Copernican planetary system. Galileo's discovery of Jupiter's moons, unknown to Copernicus, has been included.

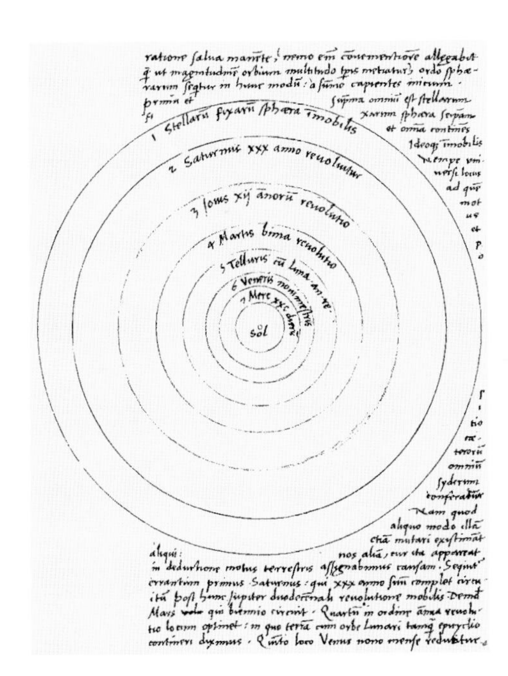

▲ A page from Copernicus' *De revolutionibus* (1543) showing the Sun in the centre and the planets in the correct order, with fixed stars in the outer shell. The epicycles Copernicus required to 'save the phenomenon' are not illustrated here.

a heliocentric (sun-centred) universe. Copernicus was educated at the prestigious University of Cracow, and also studied in Italy before taking up his position as Canon of Frauenberg, a small town on the Baltic coast. Copernicus' own system was actually not very different to Ptolemy's in that the orbits were still constructed with circular cycles and epicycles. However, placing the Sun at the centre initially simplified the number of cycles needed, though as he refined the model it came to have even more epicycles than the Ptolemaic system. The Copernican system also correctly predicted the order of the planetary orbits from the Sun, and made it possible to estimate the relative distances of each planet from the Sun. The apparent retrograde motion of the planets was now partially explained in terms of their relative motion with respect to a moving Earth, rather than in terms of motion in epicycles with respect to a stationary Earth. His great work *De revolutionibus orbium coelestium* ('On the Revolutions of the Heavenly Spheres') was published only in 1543, the year of his death, and this with some reluctance on his part.

Copernicus has given his name to a revolution in which he appears to have played an unwilling role. The ideas set forth in *De revolutionibus* are prefigured in the *Commentariolus*, a manuscript written in the early the 1510s and distributed privately. His aim seems to be have been not to overhaul the Ptolemaic system but to make it more perfect, more Greek! His quibble was that Ptolemy's model required the planets to move at variable speed along the epicycles, whereas Copernicus demanded strict adherence to Aristotelian perfect motion at constant speed about perfect circles. It was these demands that led him to make certain assumptions which, from our perspective nearly five hundred years later, appear to make him very modern. These assumptions included placing the Sun at the centre of the universe and making the Earth revolve around the Sun as well as rotate about its own axis. This heliocentric prototype was, however, no less cumbersome than the Ptolemaic system, having 34 epicycles as opposed to 40, and this for seven heavenly bodies plus the celestial sphere. The *Commentariolus* was merely an outline which Copernicus promised to flesh out later. But he seemed increasingly reluctant to actually publish, despite support from the Church authorities and the Vatican itself.

In 1514 Copernicus was invited to participate in the Lateran Council on reforming the calendar, but refused on the grounds that the calendar could not be properly reformed until the motions of the planets were securely established. Ultimately, he was unsure of the whole edifice he had constructed because he had not come up with any real proof that his system was any better or more accurate than Ptolemy's. He had relied on the astronomical tables of the ancients and seemed to have made few observations of his own. It was only the enthusiasm and efforts of his new-found admirer, Rheticus, that led to *De revolutionibus* being published – in Nuremberg, by then a Lutheran city. However, shortly before the book appeared, Rheticus moved from the University of Wittenberg to Leipzig, and the printing was entrusted to Andreas Osiander, one of the co-founders of Lutheranism. This is when the famous Preface was inserted, most probably by Osiander himself. The Preface was essentially a warning to the reader: the truth or otherwise of the Copernican system is not a serious issue: a comparison between different systems is useful in judging which one is easier to use in making calculations; and the actual heavenly motions were to be decided by other, philosophical and theological criteria. To be fair, Copernicus himself had held such doubts, but the Preface was more likely to have been inserted to appease Martin Luther, who was very much against the Copernican view, rather than the Vatican, which seemed to have encouraged Copernicus' speculations. It must also be remembered that Copernicus' work was not put on the Vatican's list of heretical works until the Counter-Reformation was well established, some eighty years after its publication.

The *Commentariolus* had made bold assertions which *De revolutionibus* largely failed to achieve. In the final version Copernicus had even more epicycles than Ptolemy, and the planets now revolved not about the Sun, but about points removed from the Sun (unwittingly, he was in a sense anticipating the true nature of planetary orbits, in which a planet followed an elliptical orbit, with the Sun at one of the foci of the ellipse, not the ellipse's centre). The one thing in its favour was that the apparent retrograde motion of the planets was shown to be a consequence of the relative motions of the planets and the Earth. The book was spectacularly unsuccessful. At the time, terrestrial motion and celestial motion were two completely different categories of phenomena. However, one belief that Copernicus did hold as being really true was that the Earth does move – his tragedy was that he had not convinced himself of how. Copernicus' name remained on people's lips thanks to the publication of his astronomical tables in 1551. *De revolutionibus* sank without trace.

'My aim is to show that the heavenly machine is not a kind of divine, live being, but a kind of clockwork (and he who believes that a clock has a soul, attributes the maker's glory to the works), insofar as nearly all the manifold motions are caused by a most simple, magnetic, and natural force, just as all the motions of the clock are caused by a simple weight. And I also show how these physical causes are to be given numerical and geometrical expression.'

Johannes Kepler, Letter to Herwart, 10 February 1605

Our reticent canon had unwittingly lit a slow-burning fuse, and there were to be explosive repercussions. Johannes Kepler (1571–1630), an ardent Copernican, was outraged by the anonymous Preface, which gave the unwary reader the impression that it was written by Copernicus himself. But Kepler himself had the courage to rebel against the tyranny of Greek astronomy. He had had a wretched childhood and was of delicate health, but his intellectual brilliance shone through and the newly established Protestant

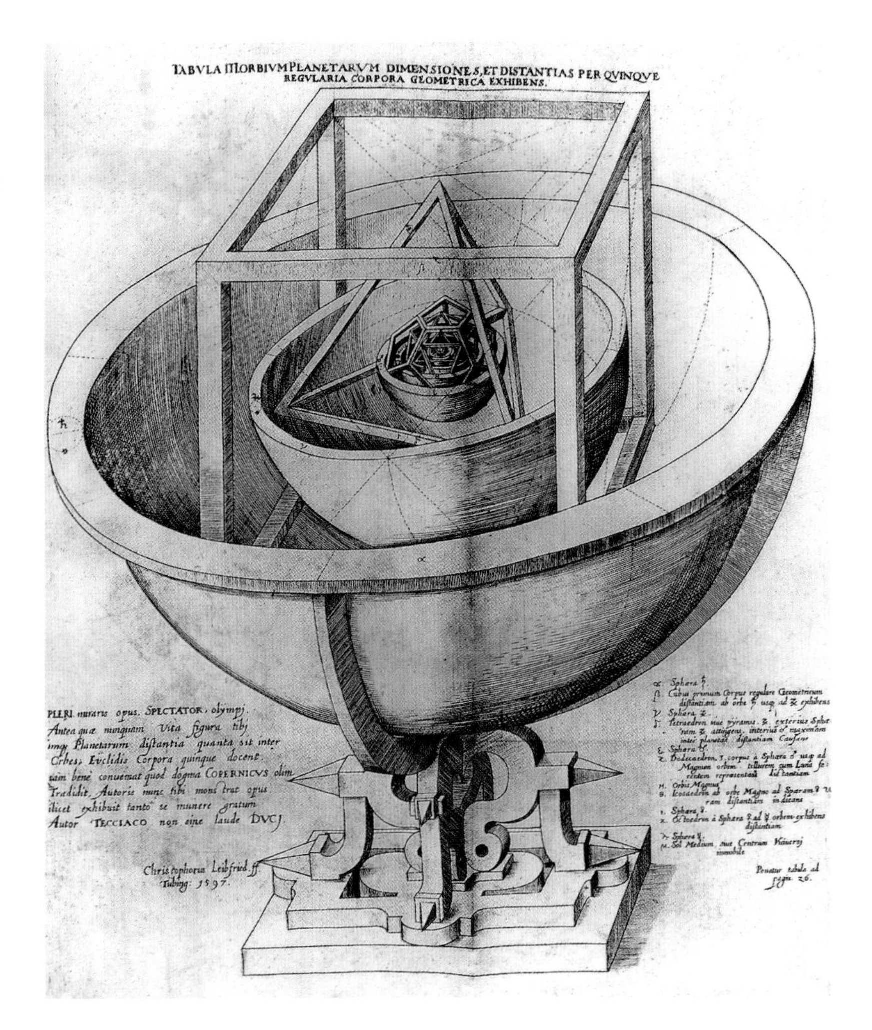

➤ Kepler's model of nested Platonic solids from the *Mysterium cosmographicum* (1596) was his first attempt to explain the relative distances between the planets. The outermost sphere denotes the planet Saturn, within which is constructed a cube. A sphere then constructed within the cube gives the orbit of Jupiter, and so on down to the orbit of Mercury.

state supported his education. He wanted to become a priest, but the governors of the Theological faculty at the University of Tübingen were obviously a little more perceptive and assigned him the position of Mathematicus in Graz as soon as it became available. Kepler represents a transitional phase in scientific understanding. His opinions on astrology changed during his lifetime: he was in no doubt that the planets had some influence on the spirit, but he was unsure precisely how. His works are an astonishing testament to the developing ideas of a scientist with all the dead ends left in.

In 1595, while teaching to a class, Kepler had his first vision of cosmic harmony. On the blackboard he was drawing a figure which consisted of an equilateral triangle with a circle inscribed within it and another circle circumscribed around it. It suddenly struck him that the ratios of the two circles were the same as the ratios of the orbits of Saturn and Jupiter as then known. This flash of inspiration led him to his famous model of nested Platonic solids. It had been known from Euclid's days that there were only five perfect solids, and there were now six known planets (including the Earth and excluding the Sun and Moon). For each solid one could construct a circumscribed sphere which just touched each vertex of the solid, and also an inscribed sphere which just touched the centres of each face of the solid. If Kepler could get the order of the solids correct, he could pack them together like a Russian doll and the spheres would correspond to the orbits of the planets. He was intoxicated with the idea and how it married mathematical precision with cosmic harmony. He published his results in the *Mysterium cosmographicum* in 1596, when he was aged twenty-five. In his

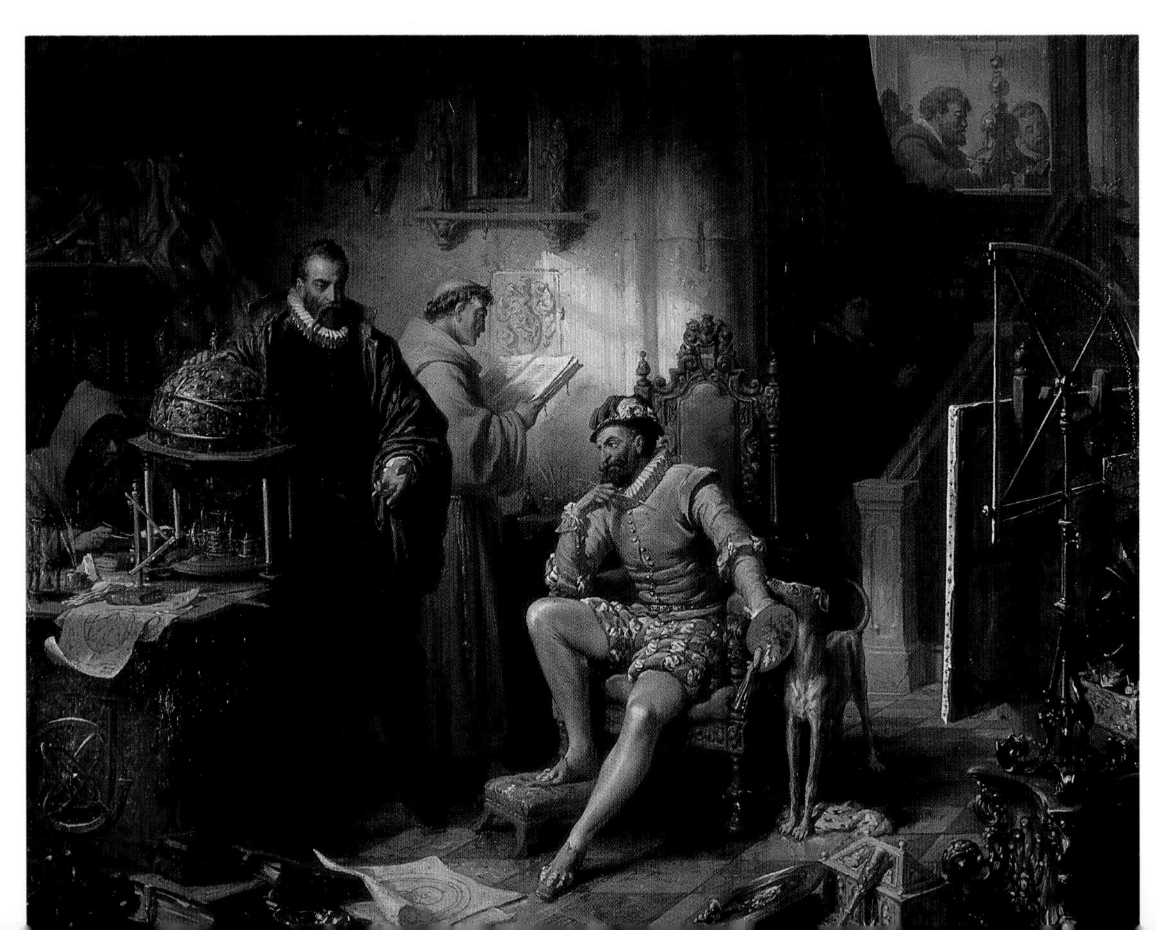

introduction he gave his first public support for the heliocentric system, and thus began Copernicus' posthumous fame. Although Kepler heeded advice not to devote a whole chapter to reconciling Copernicanism with the Scriptures, he did state in this work that the heliocentric universe was absolutely and physically true. He believed not that the solids themselves actually existed in some sense, but that the underlying structure was a sign of the Great Architect himself. After further metaphysical speculations on such topics as the Pythagorean harmony of the spheres, the *Mysterium cosmographicum* suddenly changes register and begins to read like a piece of modern mathematical physics. We see Kepler describing all the calculations and ratiocinations he had gone through. For example, Saturn is twice as far from the Sun as Jupiter is, but it takes two and half times as long to complete one revolution. Saturn is thus not only farther away but travels more slowly. Kepler searches for a physical solution, dismissing the possibility that angels are somehow more tired the farther they are from the Sun. We find here the first discussions of a kind of gravitational force emanating from the Sun which decreases in strength with distance. The source of this force was God himself, in the form of the Father emanating the Holy Ghost throughout the universe. The Creator who had previously been banished to the hypercelestial realm was now at the heart of the Solar System. The end of the *Mysterium cosmographicum* sees a return to astrological matters, with Kepler charting a horoscope for the day of Creation, Sunday 27 April 4977 BC. This was a flawed masterpiece: the theory of nested solids was false, and Kepler's version of gravitation did not work. Kepler was well aware of this but, convinced he was close to the truth, went about experimenting.

What Kepler needed was an accurate table of astronomical observations and there was one man who had one – Tycho Brahe. On receiving Kepler's book, Tycho recognized the young man's genius, and three years later Kepler was in Prague as Tycho's assistant. The two men could not have been more different. Tycho, with his famous golden nose replacing the part lost in a duel, was a flamboyant figure determined to establish an accurate knowledge of the skies; Kepler was obsessed with a mystical physics. Tycho had the best observatory anywhere and held the data that Kepler needed. However, Tycho had a planetary theory of his own, and had not only refused to publish it but had also withheld most of it from his colleagues and assistants. Tycho had been awestruck by an eclipse in his youth, but more than that he was more fascinated by the fact it had actually been predicted. In 1600 the two finally met and Kepler was assigned the task of looking at the data for Mars, notoriously the most difficult planetary orbit. Relations were always strained between the two and Tycho, the experimenter near the end of his days, knew he must bequeath his life's work to the younger Kepler so that he could design the new universe. Each needed the other. After only eighteen months, Tycho died and Kepler was made the new Imperial Mathematicus under Rudolph II.

◀ *Tycho Brahe and Rudolph II in Prague*, Eduard Ender, 1855. Tycho is demonstrating the use of a celestial globe. At the beginning of the seventeenth century, Tycho's observatory 'the island of Venus, vulgarly named Hveen' had the most accurate measurements of the heavens in the world, data that Kepler was to transform into the theory of elliptical orbits.

The observational data were now his, but turning the numbers into orbits took longer. In 1609 Kepler published his magnum opus, the *Astronomia nova*. Like his earlier work, this is less a textbook than a diary recording every twist and turn of his creative intelligence – the reader hears every whoop of joy and every cry of despair as Kepler goes to war with Mars. The difficulty with the Martian orbit is that it is the most elliptical and hence deviates the most from a circular orbit. However, it was to provide the key to all the other orbits. Kepler could not stack epicycle upon epicycle like astronomers before him. His task was not to 'save the phenomenon' but to find the laws of planetary motion and express them in the language of geometry. The triumph of the *Astronomia nova* was to establish that the planetary orbits are elliptical with the Sun at one of the two foci of the ellipse – this is the first of Kepler's laws (the word 'focus', Latin for 'fire', was first used by Kepler in this sense). The choreographed planetary pirouettes of ancient astronomy were replaced by elegant ellipses. Here too Kepler presented the second of the laws named after him: that equal areas are swept out by a planet in equal time. He also comes desperately close to the theory of gravitation modelled on the mutual attraction of magnetism, correctly ascribing tides to the attraction of the Moon, and recognizing that it is the very same gravitational force that keeps the Earth's seas from drifting off into space. But he does not develop the inverse-square law, even though he was aware that the intensity of light followed just such a law; this would be left to Newton. And, having discovered the true orbits of the planets, he was uneasy about the real motive forces behind such motions, and never did find out why the orbits are indeed ellipses. But now banished from astronomy were the unseen angels and unmoved movers. This was a universe of geometry and forces.

In 1618 Kepler returned to the leitmotif of his life with the publication of *Harmonice mundi*, a fusion of mathematics, physics and mysticism – the ultimate Pythagorean dream. Here we find Kepler's third law of planetary motion: that the square of the period of revolution is proportional to the cube of the mean distance from the Sun. The three laws together have hidden within them the theory of gravitation whose explicit pronouncement eluded Kepler. Kepler's *Epitome astronomiae Copernicanae* (1618–21), a full exposition of Keplerian astronomy, not only for Mars but for all the known planets, was essentially the most important astronomical treatise since Ptolemy's *Almagest*. But Kepler was at least a generation ahead of his contemporaries, who continued to profess the Ptolemaic doctrines. Even Galileo's *Dialogue Concerning the Two Chief World Systems* still had cycles and epicycles.

It seems that Kepler and Galileo never met face to face, despite being contemporaries. In 1597 Kepler sent a copy of his *Mysterium cosmographicum* to Galileo. At the time Galileo expressed unease at publicly supporting Copernican views. His general treatment of Kepler was at best unkind and at worst downright malicious: he feigned Kepler's friendship while at the same time refusing to send him a new telescope and

▲ *Astronomers Observing an Eclipse*, by Antoine Caron (1552–99), probably the eclipse of 1559 which had so impressed the young Tycho.

even copies of his own works, preferring to ingratiate himself with potential patrons rather than his scientific colleague. In 1609 Galileo commenced his famous observations with the newly invented telescope, one of which he presented to the Venetian Senate, which doubled his salary and made him a lifelong professor at Padua. Within a year Galileo had increased the power of the telescope and unleashed into the world his *Sidereus nuncius* ('Messenger from the Stars'). Galileo's observations revealed that the Moon was not a perfect, smooth-surfaced sphere but was covered with mountains, the planet Venus went through phases just like the Moon, and Jupiter had its own system of satellites. He even thought Saturn was a triple planet because through his fairly crude telescope its rings appeared to be two bulges flanking the planet's disc. Galileo became mathematician to the Medicis. He was honoured in Rome by election to the Accademia dei Lincei, the first scientific society in the world, and feted by the Jesuits. He had become a media star, and as he preferred to write in the vernacular rather than in Latin, his works became widely known in Italy.

The Church was concerned because the Copernican system appeared to contradict established interpretations of Scripture, but the Jesuits were prepared to accept the truth of the heliocentric system if incontrovertible proof could be found. It would not have been the first time that doctrine had been changed in the light of scientific facts, such as the spherical nature of the Earth. The Jesuits had verified all of Galileo's observations and were still supporting Kepler's work. Many books have been written on the tragedy that followed, so I shall describe it only briefly here. The Church accepted that Kepler's system worked and 'saved the phenomenon' more accurately than Ptolemy's could, but there was yet no good reason to believe in the physical reality of this extraordinary planetary system. In order to overturn a centuries-old world-view and set into motion the re-education of the lay masses to accept the new world-view, there needed to be more proof. A rearguard action against such change came from the numerous and powerful Aristotelian theologians whom Galileo had unwisely taunted with sarcastic and inflammatory language. An arrogant self-publicist, Galileo courted wealth and prestige, but when that support was removed he had few friends left in academic circles. In 1616 Galileo had pledged not to discuss Copernicanism in any way whatsoever, but in 1632 he blatantly contravened this by publishing his *Dialogue Concerning the Two Chief World Systems*. This was virtually a Copernican manifesto, and contained very thinly veiled attacks on some of the most powerful theologians of the day. The Vatican's patience had run out, and Galileo was immediately summoned to Rome. By the following year he had recanted his views and been placed under virtual house arrest. He continued to live fairly comfortably, entertaining numerous visitors, but was prohibited from ever publishing or teaching again – contemporary accounts say he was a broken man. He misjudged both his influence and the change in the prevailing mood. This was now the time of the Counter-Reformation and the Inquisition, and Europe's religious schism was brutal on both sides. Kepler spent years defending his own mother from accusations of witchcraft, and had left Prague for Austria as the Thirty Years War began. Copernicus and Kepler had worked in relative freedom, able to write what they wanted so long as they did not challenge religious authority. The Jesuit Collegium Romanum tried to reconcile scientific freedoms in a period when the Jesuit Order as a whole was spearheading the Inquisition. The whole power invested in the Papacy and the Vatican had been given a metaphysical legitimacy through a hierarchic model of the universe. This power was not only being threatened by the Reformation but also by the new physics; the suppression of the Copernican system was not out of ignorance but out of expediency. This is illustrated by the fact that soon after Galileo's trial, the Jesuits were teaching the Copernican system to impress peoples in far-off lands such as China and Japan with its predictive power.

philosophy is written in this grand book. I mean the universe which stands continually open to our gaze. But it cannot be understood unless one first learns to comprehend the language and interpret the characters in which it is written. It is written in the language of mathematics and its characters are triangles, circles and other geometrical figures, without which it is humanly impossible to understand a single word of it; without these one is wandering about in a dark labyrinth.

Galileo, *The Assayer*, 1623

During his last years Galileo still managed to write *Discourses and Mathematical Demonstrations Concerning Two New Sciences*, which was smuggled out of Italy and printed in Leiden. Here he returns to the subject of mechanics, which had been his initial inspiration, and his analysis of acceleration. His youthful analysis of the pendulum swing had shown that the time taken for each oscillation was independent of both its amplitude and the weight of the bob, and depended only on the length of the pendulum, being proportional to the reciprocal of its square root. His experiments with rolling bodies down various planes and in free fall led him to two important discoveries: that the velocity of a body was proportional to the time for which it was in motion, and that the distance travelled was proportional to the square of the time. It had also been believed that a heavier body would fall faster than a light one, but Galileo showed this to be false, stating that they fell at the same rate so long as we ignore air resistance. In practice, a cannonball will fall faster than a feather, but this is not a result of the difference in their weights, but of the different effects of air resistance – a small ball the same weight as the feather would fall at the same rate as the cannonball. Galileo had distinguished two different forces, and this was also to lead to his analysis of a projectile in flight. By separating the horizontal and vertical components he discovered that the path of a projectile is a parabola. This led to further work on ballistics for the military.

So much for the authority of Holy Scripture. Now as regards the opinions of the saints about matters of nature, I answer in one word, that in theology the weight of Authority, but in philosophy the weight of Reason alone is valid. Therefore a saint was Lactantius, who denied the earth's rotundity; a saint was Augustine, who admitted the rotundity, but denied the antipodes exist. Sacred is the Holy Office of our day, which admits the smallness of the earth but denies its motion: but to me more sacred than all these is Truth, when I, with all respect for the doctors of the Church, demonstrate from philosophy that the earth is round, circumhabited by antipodes, of a most insignificant smallness, and a swift wanderer among the stars.

Kepler, *Astronomia nova*, Introduction, 1609

Born in the year of Galileo's death, Isaac Newton was to bring together all these disparate elements into one unified theory. To understand the confusion that reigned at the time, it should be borne in mind that there were still two sciences of mechanics, terrestrial and celestial. For Kepler, the planets moved in elliptical orbits, driven by a mysterious magnetic force emanating from the Sun, with the inertia of planets slowing them down with respect to the rate of rotation of the Sun itself. For Galileo, the planets moved in circles because such motion is inherent and perfect, and inertia actually kept the planets in motion. The scene was further confused when Descartes announced, in an elaboration of Kepler's model, that inertia made bodies travel in a straight line, and that planets' paths were curved by vortices in the Solar System. Galileo's pioneering work on acceleration and terrestrial mechanics appeared to have nothing to do with celestial mechanics. There was no agreement on the definitions of key physical ideas such as mass and weight, inertia and momentum, force and energy, magnetism and gravity.

In 1687, after much pleading and financial support from Edmond Halley, Newton published his *Philosophiae naturalis principia mathematica*, or universally known simply as the *Principia*. It was not till the 1720s, after two further editions of the book had appeared, that it became widely recognized. For the purposes of this chapter I shall mention just the mechanics in the *Principia*, leaving the calculus for the next chapter. The *Principia* includes Newton's famous three laws of motion. In the traditional order, (though not the order they appeared in), the first law states that 'every body continues in its state of rest, or of uniform motion in a right line, unless it is compelled to change that state by forces impressed upon it' – thus supporting Descartes and allowing for both static and dynamic equilibrium of forces. The second law states that 'the change in motion is proportional to the motive force impressed, and is made in the direction of the

➤ Galileo's drawings of Jupiter's satellites, which he named the Medicean Stars in honour of his patrons, as illustrated in his *Sidereus nuncius* (1610). In an age when the geocentric system still held sway, Galileo's discovery was proof that another planet could act as the centre of an orbit.

right line in which that force is impressed', now expressed as $F = ma$. And the third law is that 'the mutual actions of two bodies upon each other are always equal, and directed to contrary parts'. Newton discusses various types of force field, including the inverse-square law for gravitation. His master stroke was the equating of Kepler's and Galileo's forces. In Book III of the *Principia*, entitled 'The System of the World', are the key passages where he equates the force acting on a falling body with the force acting on the planets in orbit. At a stroke, the two sciences had become one – terrestrial and celestial mechanics followed the same laws. The invisible glue that held everything together was still this mysterious gravitational force.

Newton was famed for inventing, or co-inventing, the calculus, but the proofs in the *Principia* are all geometric, though the diagrams do often represent infinitesimal changes in force and motion, showing that the resulting motion should be regarded as smooth, rather than as a series of abrupt, staccato changes. There were still some unresolved problems in Newton's cosmology. There was no obvious reason why the planets should all revolve in the same direction, nor why they sat in those precise orbits rather than others. As to the actual reality of gravitation, Newton himself was perturbed about such a powerful force acting over large distances without some medium. He did not believe in action at a distance across the vacuum of space, but rather that there was a medium, an aether, through which the force was transmitted – though whether this ether was itself material was left unresolved. The vision of angels pushing the planets had been replaced by a universal spirit. Also, if gravitation was so all-pervading, then all objects would tend to attract one another, and the whole universe would collapse. Even Newton brought in God as a kind of protector of the universe against this doomsday force. The whole theory of gravitation would have been easily dismissed were it not for the fact that the mathematical model of gravitation fitted the observed facts: the physical reality matched the analysis of the physical assumptions. Descartes's vortex system was eventually dismissed because gravitation worked better. The mathematics did more than merely save the phenomenon. The new mechanics came with a new branch of mathematics – calculus. We shall now look at the story behind its invention.

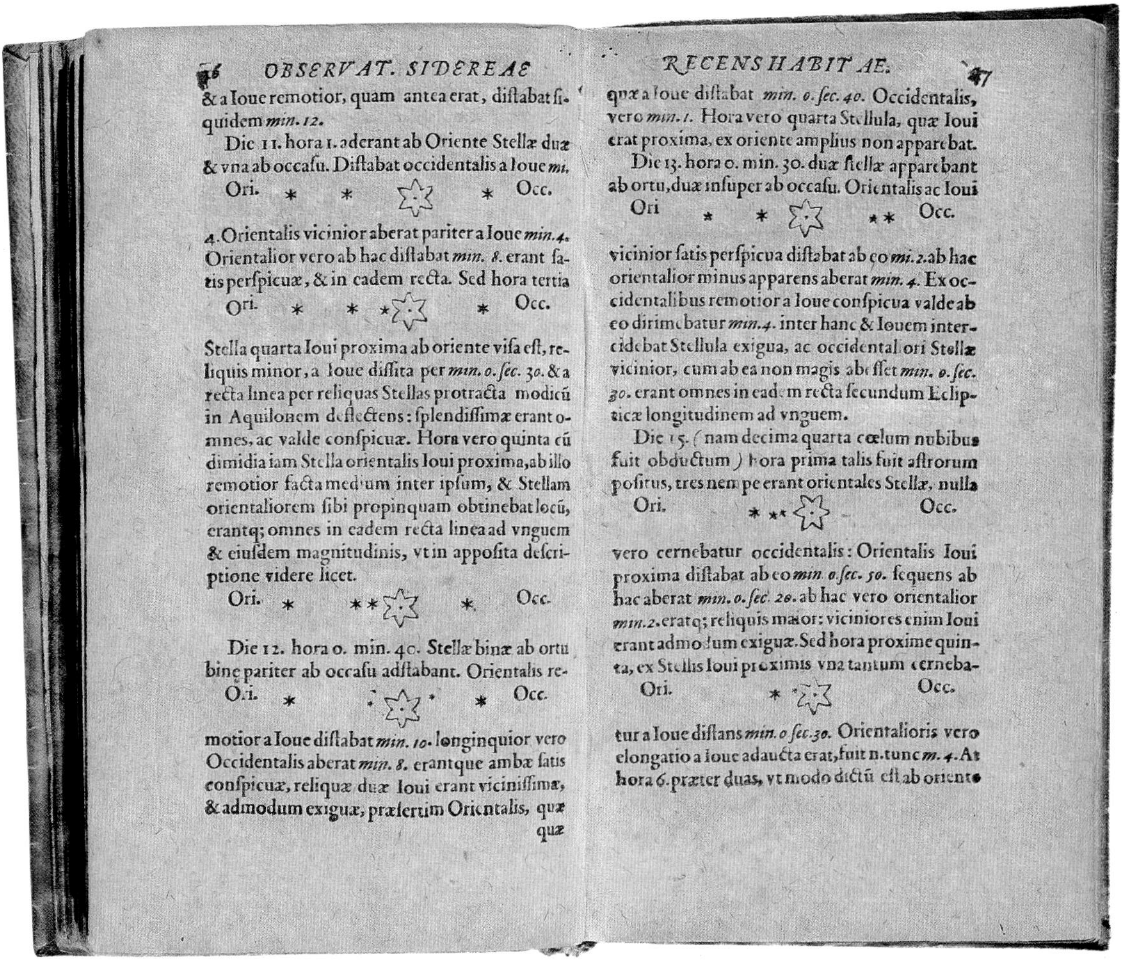

& a Ioue remotior, quam antea erat, diſtabat ſiquidem *min. 12.*

Die 11. hora 1. aderant ab Oriente Stellæ duæ & vna ab occaſu. Diſtabat occidentalis a Ioue *mi.*

Ori. Occ.

4. Orientalis vicinior aberat pariter a Ioue *min. 4.* Orientalior vero ab hac diſtabat *min. 8.* erant ſatis perſpicuæ, & in eadem recta. Sed hora tertia

Ori. Occ.

Stella quarta Ioui proxima ab oriente viſa eſt, reliquis minor, a Ioue diſſita per *min. 0. ſec. 30.* & a recta linea per reliquas Stellas protracta modicũ in Aquilonem deflectens: ſplendiſſimæ erant omnes, ac valde conſpicuæ. Hora vero quinta cũ dimidia iam Stella orientalis Ioui proxima, ab illo remotior facta medium inter ipſum, & Stellam orientaliorem ſibi propinquam obtinebat locũ, erantq; omnes in eadem recta linea ad vnguem & eiuſdem magnitudinis, vt in appoſita deſcriptione videre licet.

Ori. Occ.

Die 12. hora 0. min. 40. Stellæ binæ ab ortu binę pariter ab occaſu adſtabant. Orientalis re-

Ori. Occ.

motior a Ioue diſtabat *min. 10.* longinquior vero Occidentalis aberat *min. 8.* erantque ambæ ſatis conſpicuæ, reliquæ duæ Ioui erant viciniſſimæ, & admodum exiguæ, præſertim Orientalis, quæ

quæ

quæ a Ioue diſtabat *min. 0. ſec. 40.* Occidentalis, vero *min. 1.* Hora vero quarta Stellula, quæ Ioui erat proxima, ex oriente amplius non apparebat.

Die 13. hora 0. min. 30. duæ ſtellæ apparebant ab ortu, duæ inſuper ab occaſu. Orientalis ac Ioui

Ori Occ.

vicinior ſatis perſpicua diſtabat ab eo *mi. 2.* ab hac orientalior minus apparens aberat *min. 4.* Ex occidentalibus remotior a Ioue conſpicua valde ab eo dirimebatur *min. 4.* inter hanc & Iouem intercidebat Stellula exigua, ac occidentaliori Stellæ vicinior, cum ab ea non magis abeſſet *min. 0. ſec. 30.* erant omnes in eadem recta ſecundum Eclipticæ longitudinem ad vnguem.

Die 15. (nam decima quarta cœlum nubibus fuit obductum) hora prima talis fuit aſtrorum poſitus, tres nempe erant orientales Stellæ, nulla

Ori. Occ.

vero cernebatur occidentalis: Orientalis Ioui proxima diſtabat ab eo *min. 0. ſec. 50.* ſequens ab hac aberat *min. 0. ſec. 20.* ab hac vero orientalior *min. 2.* eratq; reliquis maior: viciniores enim Ioui erant admodum exiguæ. Sed hora proxime quinta, ex Stellis Ioui proximis vna tantum cerneba-

Ori. Occ.

tur a Ioue diſtans *min. 0. ſec. 30.* Orientalioris vero elongatio a Ioue adaucta erat, fuit n. tunc *m. 4.* At hora 6. præter duas, vt modo dictũ eſt ab oriente

I Galileo, son of the late Vincenzo Galilei, Florentine, aged seventy years, arraigned personally before this tribunal and kneeling before you, most eminent and reverend Lord Cardinals' Inquisitor General against heretical gravity throughout the entire Christian commonwealth, having before my eyes and touching with my hands the Holy Gospels, swear that I have always believed, do believe, and by God's help will in the future believe, all that is held, preached and taught by the Holy Catholic and Apostolic Church. But whereas — after an injunction had been judicially intimated to me by this Holy Office to the effect that I must altogether abandon the false opinion that the sun is the centre of the world and immovable and that the earth is not the centre of the world and moves, and that I must not hold, defend or teach in any way whatsoever verbally or in writing the said false doctrine, and after it had been notified to me that the said doctrine was contrary to Holy Scripture — I wrote and printed a book in which I discuss this new doctrine already condemned and adduce arguments of great cogency in its favour without presenting any solution of these, I have been pronounced by the Holy Office to be vehemently suspected of heresy, that is to say, of having held and believed that the sun is the centre of the world and immovable and that the Earth is not the centre and moves:

Therefore, desiring to remove from the minds of your Eminences, and of all faithful Christians, this vehement suspicion justly conceived against me, with sincere heart and unfeigned faith, I abjure, curse and detest the aforesaid errors and heresies.

Galileo's public recantation of the heliocentric theory in 1633

chapter thirteen

mathematics in motion

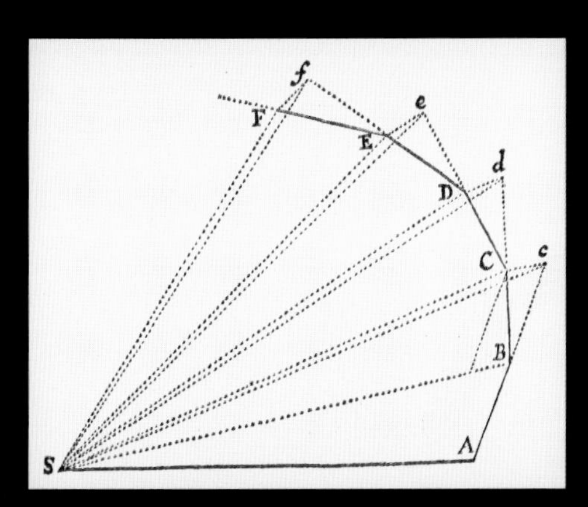

◄ Detail of Book 1, Proposition 1, Theorem 1 – Newton's *Principia* (1687).

We saw earlier that both Newton and Kepler modelled the planetary orbits in an essentially geometric manner. However, ellipses themselves have no physical existence in space – they are just the invisible paths traced out by an orbiting planet. It would therefore be very useful to find a mathematical tool which could describe the planets *in motion* rather than having to construct their paths geometrically, point by point. But those who tried to make the transition from a sequence of rectilinear motions to a truly smooth path resurrected the problems of infinities and infinitesimals.

Before looking at the invention of the calculus, it is worth considering earlier attempts to handle general problems of areas and tangents. This 'pre-calculus' can be found as early as Archimedes, who had developed two methods for finding areas bounded by curved lines, often referred to as the geometrical and mechanical methods. One of the most famous problems bequeathed by the ancients was finding the area of a circle, the so-called quadrature, or squaring of the circle. In a short treatise, *On the Measurement of Circles*, Archimedes proves two important results, firstly that the area of a circle is equal to the area of a right-angled triangle whose base is the circumference of the circle and whose height is the radius of the circle, equivalent to our formula πr^2 but without the need to express π explicitly. The second major result was a proof that the numerical value of π lies between $3\frac{1}{7}$ and $3\frac{10}{71}$. In both cases, the geometrical method used was to draw inscribed and circumscribed polygons of the circle; then, by repeated doubling of the number of sides of each polygon, they come to fit ever closer to the circumference of the circle. Not only that, but the two polygons tend to come closer to each other, in a sense sandwiching the circle, so that if the process were continued ad infinitum (what mathematicians call 'in the limit'), the areas of the polygons tend to the area of the circle. To find the value of π, Archimedes started with circumscribing and inscribing hexagons, and terminated the process once he had reached a 96-sided polygon, though he could have continued until he had reached any level of accuracy decided beforehand. The procedure was justified by employing the method of exhaustion, credited to Eudoxus (Chapter 4), but Archimedes avoids stating that the polygons somehow become the circle, clinching the proof through a lengthy logical argument. This reticence is understandable, avoiding as it does a kind of leap of faith necessary to pass from a polygon to a circle – which, to the Greeks, were very different types of object.

Archimedes' mechanical method is exemplified in *The Method*, originally a lengthy letter addressed to Eratosthenes and presumed lost but discovered at Constantinople in 1906. It was in the form of a palimpsest, a tenth-century parchment of various works by Archimedes which had later been imperfectly washed clean in order to be reused for prayers, but the original mathematical texts could still be made out. (It was auctioned in 1998 for $2 million.) The method Archimedes discusses is essentially to deconstruct an area into lines, transform the lines and reconstruct them into another area. The precise

▲ Cellarius, *Atlas Coelestis*, 1660. This lavishly illustrated work surveyed the various planetary models at the time. Within a few years Newton's *Principia* (1687) would revolutionize mathematical physics and planetary theory.

transformation was performed by using Archimedes' rules on the operation of a lever. In a sense, Archimedes was balancing a known area with an unknown area, the position of the fulcrum determining their relative sizes – hence the term mechanical'. Although Archimedes claimed this to be a very useful heuristic tool for discovering new results, he was also aware that it did not constitute a valid method of proof, and he reverted to the geometrical method when it came to presenting his polished results. The main problem lay in assuming that an area could be composed of indivisible lines, for a line is length without width, a one-dimensional entity, and however tightly we mentally bundle the lines together, the sum of one-dimensional objects remains one-dimensional, not a two-dimensional area. Despite such misgivings Archimedes succeeds in calculating correctly a number of areas and volumes, including the area of a segment of a parabola, and the centres of gravity of volumes such as the cone.

By the early seventeenth century interest had grown in generating a variety of curves and discovering their lengths, the areas under them and the volumes generated

▲ The astronomer and astrologer from Robert Fludd, *Utriusque Cosmi Historia*, (1617–24), a vast work on cosmic harmony combining the physical and spiritual sciences through the correspondences between macrocosm and microcosm.

by rotating them. The impetus came from a variety of mechanical interests in both statics and dynamics. Establishing the centre of gravity of an object mathematically was important in deciding on its stability, and this was of obvious concern in areas such as architecture and shipbuilding. The methods that were used essentially fell into the two Archimedean categories, but it was increasingly felt that, despite their logical problems, methods involving indivisibles or infinitesimals in some form yielded correct results more easily than did the geometrical methods.

Mathematics could no longer avoid grappling with the concepts of infinity and infinitesimals – the Scylla and Charybdis of Greek mathematics. Kepler had used infinitesimal methods in calculating the area swept out by a planet in an elliptical orbit. More impressively, in a book entitled *Stereometria doliorum* ('Volume Measurement of Barrels', 1615), he found the volume of a wine cask by using infinitely many infinitesimal slices. Galileo believed in the real existence of the infinite, citing as an example the circle as a polygon with an infinite number of sides. In the same period, Bonaventura Cavalieri (1598–1647), a student of Galileo and professor of mathematics at Bologna from 1629, published a hefty tome of nearly seven hundred pages on methods of finding areas and volumes. His *Geometria indivisibilibus continuorum* (1635) discussed various methods of indivisibles, manipulating areas as if they were composed of indivisible lines, and volumes as if they were composed of indivisible areas. His most general result was a formula for the area beneath curves of the form $y = x^n$ for any integer value of n.

Let us now briefly look at pre-calculus developments in finding tangents to curves. Pierre de Fermat (1601–65) developed some important results but did not publish them formally, relying instead on their dissemination via the web of mathematical correspondences orchestrated by Marin Mersenne (1588–1648). Fermat developed methods for finding tangents to any point of a polynomial curve, as well as for establishing its maxima and minima. He also rediscovered Cavalieri's rules for areas under curves of the form $y = x^n$, extending them to allow for n being both positive and negative. The only anomalous case was $n = -1$, for which it was known that the solution was the logarithmic function. The methods employed by Fermat are very close to those we still use in the differential calculus, except that Fermat did not make use of the concept of the limiting process. In none of Fermat's writings on infinitesimal analysis does he mention one of the key aspects of this new analysis, which is that the problem of tangents and of areas are essentially inverses of each other. Neither did he appear to extend the range of functions amenable to his methods.

The plethora of pre-calculus methods was soon to be moulded into a new branch of mathematics. As in so much of history, what becomes revolutionary is often already in the air, just waiting for someone to grasp it firmly and mould it into shape. In this case, the invention of calculus is credited to two men, Isaac Newton and Gottfried Leibniz. As in any shared invention, there is always the lingering doubt that one person actually got there first, and the whiff of a priority dispute was to spread across Europe.

Isaac Newton was born on Christmas Day in 1642, the year of Galileo's death. In 1661 he enrolled at Trinity College, Cambridge, graduating in 1664. During the following two years the college was closed because of the plague, and Newton returned to live at home in Lincolnshire. He later wrote that it was during this period that he made his famous discoveries about infinite series, gravitation and calculus. This would seem an oversimplification, but in 1669 he wrote the *De analysi per aequationes numero terminorum infinitas*, in which he treated infinite series of powers in the same way as finite series, and he later extended the binomial theorem to any rational power. The *De analysi* also contained the first account of calculus, based on a method similar to Fermat's but with the increased power provided by the handling of infinite series. It is also the first time that finding an area under a curve was explicitly presented as the inverse of finding the tangent. In 1671 Newton wrote another paper on what he was now calling fluxions and fluents. In this work he pictured the quantities x and y as flowing with respect to time, and $\dot{x}$ and $\dot{y}$ are their rates of change, or fluxions. The quantities of which x and y are themselves fluxions are labelled x' and y'. Newton came to these ideas by considering a line as the locus of a point travelling in space. Time serves as an invisible chronometer in the scheme and does not appear as a separate variable *t*. It is unfortunate that Newton kept all his notes to himself whilst circulating some of his works only to colleagues. *De analysi* was not published until 1711, while a description of the method of fluxions appeared in English in 1736. His first public account appears in some terse and hard-to-follow passages in the *Principia* in 1687. The *Principia* itself seems to be largely free of the calculus, Newton casting all his mathematical physics in geometric terms. His stubborn refusal to publish may well be explained by an aversion to the public controversies and disputations that might have ensued, as had happened between him and Robert Hooke over the subject of optics (Newton waited for Hooke's death before publishing his *Opticks*). Even the *Principia* would never had seen the light of day had it not been for the encouragement and financial support of Edmond Halley. For whatever personal reasons Newton just wanted to be left alone to work. This was to lead to his most acrimonious battle.

In the *Principia* there is a section entitled 'The method of first and last ratios of quantities' which gives geometric demonstrations of his key ideas on the differential and integral calculus, and another section lists some results of what he calls the 'moment of any genitum', which we would now call the differential of a term. This being the first public expression of the new calculus, it is no surprise that apart from a few mathematicians the scientific establishment was initially underwhelmed. Newton went from geometric proofs to general results without going through the algebraic manipulations. He admitted in the text that such a demonstration may be easier to present, but he was still concerned that proof by indivisibles rested on shaky foundations. Newton was not the first person to handle differentiation and integration, but he was the first to create a solid framework in which the two operations were inverses of each other, and by his work on infinite series he expanded enormously the range of functions that could be handled.

And what are these fluxions? The velocities of evanescent increments. And what are these evanescent increments? They are neither finite quantities, nor quantities infinitely small, nor yet nothing. May we not call them the ghosts of departed quantities?

Bishop Berkeley, *The Analyst*, 1734

Let us take a closer look at the problem that Newton was grappling with. If we take a point on a curve and wish to find the slope of the tangent at that point, we can pick a second point close to the first and join the two with an extended line. We can also construct a right-angled triangle in which these two points are at the ends of the hypotenuse. The ratio of the other two sides of the triangle gives us the slope of the line joining the points. If we imagine the second point slowly moving towards the first, we can see that the slope gets closer and closer to being the tangent, while at the same time our triangle gets smaller and smaller. If we imagine the two points meeting, we would be confident that the tangent had been established, but the triangle has disappeared and the two sides that provided us with the

➤ Book 1, Proposition 1, Theorem 1, from Newton's *Principia* (1687), showing the path taken by a particle under the influence of a centripetal force from a fixed point. Newton proves that the area swept out by such a particle is proportional to the time taken, thereby generalizing Kepler's second law.

[37]

SECT. II.

De Inventione Virium Centripetarum.

Prop. I. Theorema. I.

Areas quas corpora in gyros acta radiis ad immobile centrum virium ductis deſcribunt, & in planis immobilibus conſiſtere, & eſſe temporibus proportionales.

Dividatur tempus in partes æquales, & prima temporis parte deſcribat corpus vi inſita rectam *AB*. Idem ſecunda temporis parte, ſi nil impediret, recta pergeret ad *c*, (per Leg. I) deſcribens lineam *Bc* æqualem ipſi *A*B, adeo ut radiis *AS*, B*S*, *cS* ad centrum actis, confectæ forent æquales areæ *A S*B, B*Sc*. Verum ubi corpus venit ad B, agat vis centripeta impulſu unico ſed magno, faciatq; corpus a recta B*c* deflectere & pergere in recta B*C*. Ipſi B*S* parallela agatur *cC* occurrens B*C* in

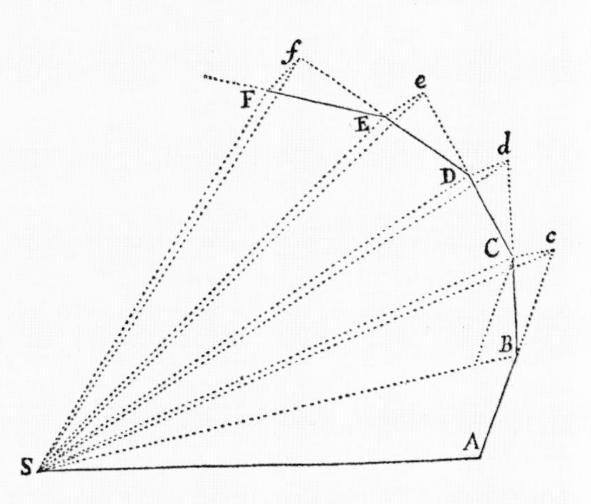

C, & completa ſecunda temporis parte, corpus (per Legum Corol. 1) reperietur in *C*, in eodem plano cum triangulo A*SB*. Junge *SC*, & triangulum *SBC*, ob parallelas SB, *Cc*, æquale erit triangulo *SBc*, atq; adeo etiam triangulo *SAB*. Simili argumento ſi vis

deſcripto, ſecetur producta recta *V R* in *H*, & umbilicis *S*, *H*, axe tranſverſo rectam *H V* æquante, deſcribatur Trajectoria. Dico factum. Namq; *V H* eſſe ad S *H* ut *V* K ad S K, atq; adeo ut axis tranſverſus Trajectoriæ deſcribendæ ad diſtantiam umbilicorum ejus, patet ex demonſtratis in Caſu ſecundo, & propterea Trajectoriam deſcriptam ejuſdem

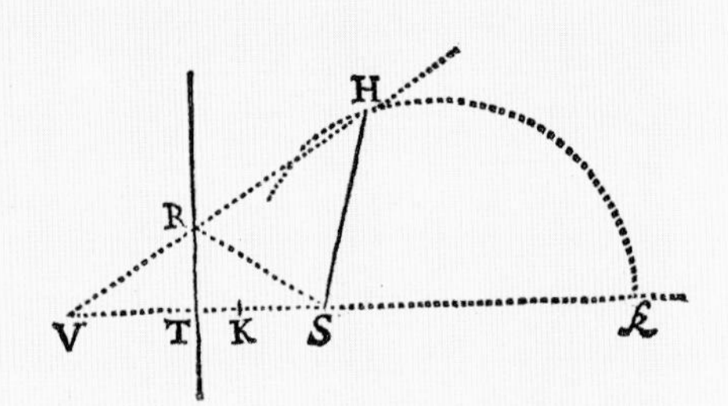

eſſe ſpeciei cum deſcribenda: rectam vero *T R* qua angulus *V R S* biſecatur, tangere Trajectoriam in puncto *R*, patet ex Conicis *Q. E. F.*

Cas. 4. Circa umbilicum S deſcribenda jam ſit Trajectoria *A P B*, quæ tangat rectam *T R*, tranſeatq; per punctum quodvis *P* extra tangentem datum, quæq; ſimilis ſit figuræ *a p b*, axe

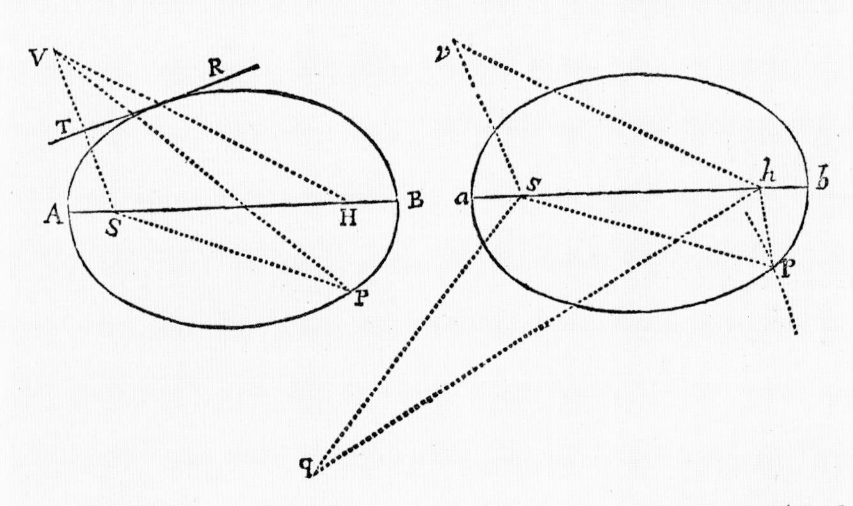

tranſverſo *a b* & umbilicis *s*, *h* deſcriptæ. In tangentem *T R* demitte perpendiculum *S T* & produc idem ad *V*, ut ſit *T V* æqualis

◄ Book 1, Proposition 20, Problem 12, from Newton's *Principia* (1687), describing various trajectories from the point of view of a focus of an ellipse. Newton had earlier shown that an ellipse, parabola and hyperbola all satisfied the inverse-square law.

numerical gradient have both vanished to zero. We are then left with a ratio of two zeros that gives us a real answer! In Newton's language, our ultimate ratio of evanescent quantities is itself a quantity. For the moment, the validity of the calculus was settled to Newton's satisfaction, and its tremendous usefulness ensured its widespread application, but the doubts about the soundness of its foundations were to persist, and the problem of the mathematics of the infinitely large and infinitely small would return. Shortly after Newton's death the philosopher Bishop Berkeley published a blistering attack on the calculus in his *The Analyst*, which although highlighting logical problems of which mathematicians were fully aware, was laced with religious bigotry, branding mathematicians as infidels for believing in the 'ghosts of departed quantities'.

Gottfried Wilhelm Leibniz (1646–1716) was born in Leipzig, where he studied theology, law, philosophy and mathematics. The university refused him a doctorate in law because he was too young at twenty, so to graduate he went to Nuremberg, where he in turn refused an offer of a professorship in law, preferring instead to become a diplomat, eventually for the Hanoverians. He is often regarded as the last great universalist, with a deep interest in logic and the foundations of a universal language. It is perhaps fitting that the language of the calculus we use today is largely Leibniz's. The terms 'differential calculus' and 'integral calculus' are his, as is the notation dy/dx and ∫ydx. His diplomatic post gave Leibniz ample opportunity to travel. He visited London in 1673, where he became a member of the Royal Society, returning in 1676 to demonstrate his new mechanical computer. He did not make Newton's acquaintance, but much of the later quarrel centred around whether Leibniz had had the opportunity to see Newton's *De analysi* during these visits. The two men were soon conducting an amicable correspondence, exchanging views on infinite series.

Although Leibniz's calculus also grew out of analyzing series, it took a rather different form: he became fascinated with summing infinite series. While in Paris he was set the problem of finding the sum of the reciprocals of the triangular numbers, represented by the general term $2/[n(n+1)]$. This he cleverly rewrote as the difference between two terms, that is, $2[1/n - 1/(n+1)]$, and just by writing out the first few terms it became obvious that all the terms cancelled out, save the first and last. Extending the sum to an infinite number of terms yielded the answer 2. Leibniz toyed with many other series, gaining experience in deciding whether they converged or diverged. He then realized that the problem of finding a tangent to a curve amounted to finding the ratio of the differences in ordinates and abscissas, the x and y values, as these became infinitely small, and that quadratures depended on the sum of the ordinates, or of infinitely narrow rectangles making up the area under the curve. Just as the sums and differences he had worked on in numerical series were inverses of each other, so the tangent and quadrature problems were also inverses. It all hinged on the characteristic infinitesimal triangle, the same triangle Newton described as the 'ratio of evanescent quantities'. Leibniz's key concept was the differential d*x* as an infinitely small change in the value of *x*. For a function $y = f(x)$, the gradient was given by dy/dx and the quadrature by $\int y dx$. The notation for the integral can almost be read as a statement that it is the sum of rectangles of sides *y* and d*x*. Leibniz's manuscripts date from 1675, and after a few changes to the notation he published his results in 1684 with a second article in 1686, both in the journal *Acta eruditorum*, which he had co-founded. We find here the standard theorems of calculus, including the fundamental theorem of calculus – that differentiation and integration are inverse processes. Leibniz stressed that the new calculus provided a universal algorithm for resolving tangent and quadrature problems for a whole range of functions, including transcendental functions, a term coined by Leibniz to denote functions, such as $\sin x$ and $\ln x$, which can be expressed as infinite power series but are not solutions of algebraic equations.

The results Leibniz obtained are similar to those Newton had obtained but withheld from publication. The dispute that erupted over priority claims for the invention of the calculus soured the last years of both men's lives. In terms of publication dates, the first edition of the *Principia* came out in 1687, after Leibniz's articles in the *Acta eruditorum*. Newton had sent a copy of the *Principia* to Leibniz, thinking he was in Hanover. Leibniz, then in Italy, read a review of the book in 1689 in the *Acta*, and based on the review wrote further papers on mechanics and optics which obviously did owe something to Newton's work. But it was the success of Leibniz's previous articles on the Continent which led many Europeans to ascribe the calculus to him. In 1699 a paper given at the Royal Society by a minor mathematician implied that Leibniz had taken his ideas from Newton. A tit for tat ensued. Leibniz employed the *Acta* as his mouthpiece, while Newton had the support of the Royal Society, which itself had to set up a committee to look into the whole matter. In 1705 the *Acta* gave an unfavourable review of Newton's latest publication, and in 1712 the Royal Society's committee decided that Newton was the first inventor. In 1726, after Leibniz's death, Newton removed from the third edition of the *Principia* any references to Leibniz which were in the original. Had Newton openly published his *De analysi* way back in 1669, the whole sorry business may have been avoided. The British retained the Newtonian calculus of fluxions and fluents until the early nineteenth century, but the development of the calculus into an incredibly powerful tool was to take place in mainland Europe and in the language of Leibniz.

For a man who had shunned publication, Newton's later years were filled with public appointments. In 1696 he was appointed Warden of the Mint, and elevated to Master in 1699 for doing such a fine job in reforming the coinage and sending counterfeiters to the gallows. In 1701 he represented Cambridge University in Parliament for a second term. In 1699 he was elected as only the second foreign associate of the Académie des Sciences, the first having been Leibniz. In 1703 he was elected President of the Royal

▲ William Blake's *Newton* (1795). 'For Bacon & Newton, sheath'd in dismal steel, their terrors hang/Like iron scourges over Albion...' William Blake, *Jerusalem*, Chapter 1.

Society, and was continually re-elected till his death. In 1705 he was knighted by Queen Anne. He is buried in Westminster Abbey, and Voltaire commented that, 'He lived honoured by his compatriots and was buried like a king who had done well by his subjects.'

Leibniz continued to develop his broad interests in philosophy, religion and a universal logic (thus presaging George Boole – see Chapter 17), having time in 1700 to help set up the Berlin Academy of Sciences, with similar plans for St Petersburg only coming to fruition after his death. In 1714 it seemed as if he was destined to live in London, his employer having become the first Hanoverian King of England. But after his sterling services as diplomat, historian, lawyer and tutor, he was asked to remain in the library researching the tangled royal family tree. Perhaps it had been suggested that it would be unwise for Newton and Leibniz to share the same court.

To end on a happier note, in 1701, in response to a request from the Queen of Prussia, Leibniz replied that 'taking Mathematicks from the beginning of the world to the time of Sir Isaac what he had done was the better half.' And in a letter written to Leibniz back in 1676, Newton says that Leibniz's 'method for obtaining convergent series is certainly very elegant, and it would have sufficiently revealed the genius of the author, even if he had written nothing else.' Mercifully, history will remember the genius of both men.

Those things which have been demonstrated of curved lines, and the surfaces which they comprehend, may be easily applied to the curved surfaces and contents of solids. These Lemmas are premised to avoid the tediousness of deducing involved demonstrations ad absurdum, according to the method of the ancient geometers. For demonstrations are shorter by the method of indivisibles; but because the hypothesis of indivisibles seems somewhat harsh, and therefore that method is reckoned less geometrical, I chose rather to reduce the demonstrations of the following propositions to the first and last sums and ratios of nascent and evanescent quantities, that is, to the limits of those sums and ratios, and so to premise, as short as I could, the demonstrations of those limits. For hereby the same thing is performed by the method of indivisibles; and now those principles being demonstrated, we may use them with greater safety. Therefore if hereafter I should happen to consider quantities as made up of particles, or should use little curved lines for right ones, I would not be understood to mean indivisibles, but evanescent divisible quantities; not the sums and ratios of determinate parts, but always the limits of sums and ratios; and that the force of such demonstrations always depends on the method laid down in the foregoing Lemmas.

Perhaps it may be objected, that there is no ultimate proportion of evanescent quantities; because the proportion, before the quantities have vanished, is not the ultimate, and when they are vanished, is none. But by the same argument it may be alleged that a body arriving at a certain place, and there stopping, has no ultimate velocity; because the velocity, before the body comes to the place, is not its ultimate velocity; when it has arrived, there is none. But the answer is easy; for by the ultimate velocity is meant that which the body is moved, neither before it arrives at its last place and the motion ceases, nor after, but at the very instant it arrives; that is, that velocity with which the body arrives at its last place, and with which the motion ceases. And in like manner, by the ultimate ratio of evanescent quantities is to be understood the ratio of the quantities not before they vanish, nor afterwards, but with which they vanish. In like manner the first ratio of nascent quantities is that with which they begin to be. And the first and last sum is that with which they begin and cease to be.

Newton's *Principia*, Scholium, Section 1, Book 1, 1726

chapter fourteen

oceans and stars

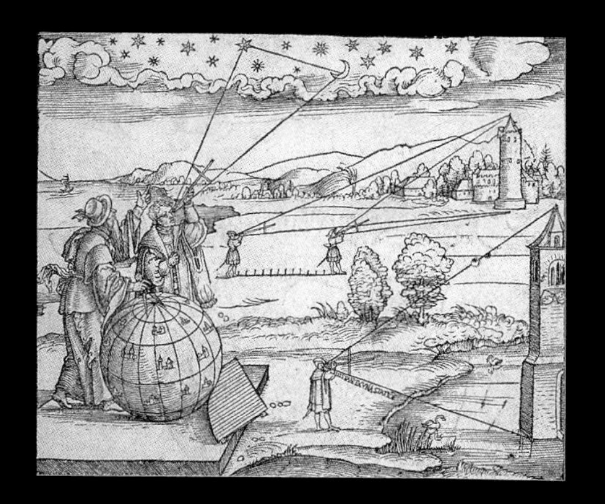

◀ Petrus Apianus, *Introductio geographica*, 1533, based upon Ptolemy's *Geography*. This image illustrates the various uses of the cross staff for finding both celestial and terrestial distances.

Map-making was undertaken by all the early civilizations. Whether for the purposes of building, taxation or planning for war, the work of the surveyor is one of the oldest professions involving practical mathematics. Dated at about 2200 BC, one of the statues representing Gudea, ruler of the Sumerian city-state of Lagash, shows the surveyor holding a scale plan of the Ningirsu temple, along with a measuring ruler and a writing implement. This is the earliest known example of a plan drawn to scale. Maps of the known world are found on Babylonian clay tablets, Egyptian papyri and Chinese silks. The Romans carried on the Greek traditions of mensuration, their *Corpus agrimensorum* laying down rules for surveying and the drawing of scale maps.

In mapping a small region we can be fairly safe in assuming that the region is flat, but as we seek to represent larger regions, the curvature of the Earth becomes a significant factor. It is not clear exactly when humans came to the realization that the Earth was spherical; in some traditions only one hemisphere was assumed to be inhabited. Eratosthenes, the head librarian in Alexandria from 240 BC, made the first known map on scientific principles, with parallels and meridians on an irregular grid. His contemporaries seemed unimpressed, and it was Claudius Ptolemy's *Geography*, in about AD 150, which became the standard work on cartography. It states that the Earth is spherical, but only partly inhabited, and gives a circumference for the Earth of 180,000 stadia, which is

▶ French Carte Pisane of *c.*1290, this is the oldest existing portolan chart showing navigation routes in Europe and the Mediterranean.

➤ The Mediterranean and North Africa from a world map made in 1500 by Juan de la Cosa, who sailed with Christopher Columbus in 1492.

actually far less accurate than Eratosthenes' value of 250,000 stadia (one stadium is believed to be equivalent to about 160 metres). The great contribution of the *Geography* was in laying the foundations for projecting a sphere onto a flat surface. Ptolemy's map was updated by al-Khwarizmi (see Chapter 7), who relied on Ptolemy's knowledge of the Mediterranean lands but improved its accuracy for Central Asia.

The projection of the spherical Earth onto a flat map will always introduce some distortion, and the map-maker's prime concerns determine which factors are distorted the most and which are distorted the least. Conformal projections minimize distortions of angles and shapes, equal-area projections preserve the relative areas, and equidistant projections preserve distances. And as we shall see, there are also different requirements for maps of land masses and charts of the seas.

With the growth of European navigation and trade from the 1300s, we find portulan charts (from the Italian *portolano*, which originally referred to written sailing directions) that bore a network of straight lines, or rhumb lines, to assist navigators in planning voyages round Europe and the Mediterranean. These were mainly produced in Venice, Genoa and Majorca, and were remarkably accurate, even though it is not clear whether they were constructed with any particular projection in mind. The extent of the use of the compass, a Chinese invention, and the state of astronomical navigation in this period are still subject to speculation. But with the discovery of America and the first printing of Ptolemy's *Geography*, the scene was set for the piecing together of an accurate map of

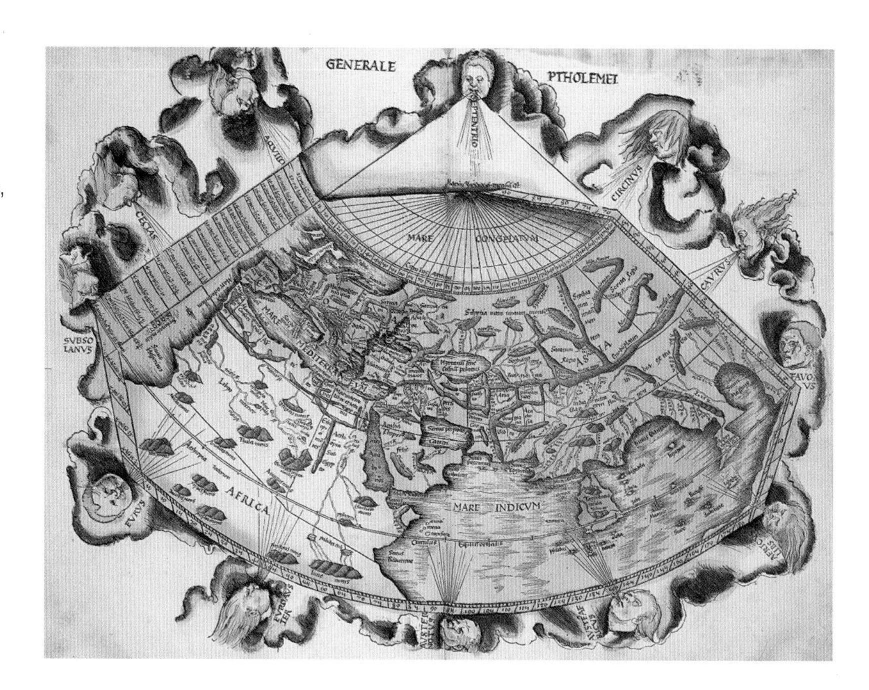

➤ A map of the world of 1513 according to Ptolemy's *Geography*, which had only recently been rediscovered in Europe.

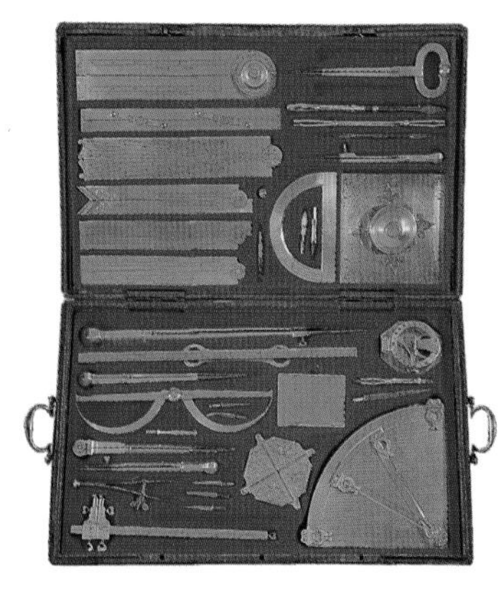

▼ A set of mathematical instruments, made in 1701 by D. Lusuerg of Rome. Such an ornate collection, including a geometric quadrant, a universal sundial and a set of Napier's rods, would probably have been made for a wealthy client as a status symbol rather than for practical use.

the world. Ptolemy's *Geography* resurfaced in Europe as late as the fifteenth century, and was first printed in Bologna in 1477. During the Renaissance various projections were employed, often for aesthetic reasons, such as the popular oval world map first used by Francesco Rosselli in 1508. These projections were based on graphical constructions rather than trigonometric formulas.

Gerardus Mercator (1512–94), described as the 'Ptolemy of our Age', designed the first projection specifically to aid seafarers. Mercator studied at the University of Louvain, graduating in philosophy and continuing his studies with mathematics, astronomy and cartography. He also became a master engraver and instrument-maker. From the mid-1530s he made a number of maps, including ones of Flanders and Palestine. Imprisoned for heresy in 1544, he was released early after strenuous lobbying from the University, but then moved to Duisburg in the Duchy of Cleve, now in Germany, and became Court Cosmographer to Duke Wilhelm in 1564. It was in Duisburg in 1569 that he designed the famous Mercator projection for his map of the world. The novelty of this projection was that it shows lines of constant bearing as straight lines, which made practical charts much easier to use by navigators. On a sphere, if a ship is set to travel on a fixed bearing (unless it be towards one of the cardinal points) its general path will be a curve on the sphere; in fact, if it were possible to travel on a fixed bearing continuously one would spiral towards one of the poles. But projecting these rhumb lines onto straight lines considerably eases the navigator's task. Another advantage is that the Mercator projection preserves angles, so that changing course by, say, 30° means that the new rhumb line makes an angle of 30° with the previous course. It has since become the most popular projection for maps of the world, although it does greatly distort shapes at higher latitudes, and some would like to see it replaced by an equal-area projection such as the recent one named after Arno Peters.

The mathematical analysis of the Mercator projection was first given by Edward Wright in *Certaine Errors in Navigation* (1599), and the same year saw the publication of Wright's world map, based on the Mercator projection, in Richard Hakluyt's *Principal Navigations*. The increased knowledge of both the terrestrial and the heavenly spheres made popular the production of twin globes – often for teaching purposes, but also as symbols of the new knowledge – in which a terrestrial globe was often encased in a hinged celestial globe. With the increased accuracy of astronomical observations, and the beginnings of the great triangulation projects in France, Britain and other European countries, world maps were in need of regular updating.

But in order to create accurate maps and charts, accurate latitudes and longitudes of key places were required. Latitude had always been fairly simple to deduce – it was exactly the same as the altitude of the celestial pole. During the day, one could use the position of the Sun, adjusted with the use of declination tables, which gave the Sun's

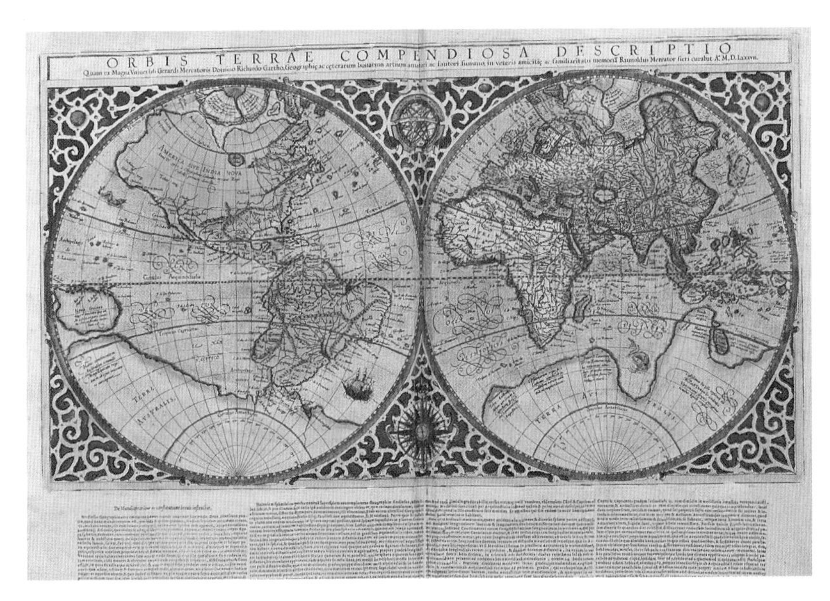

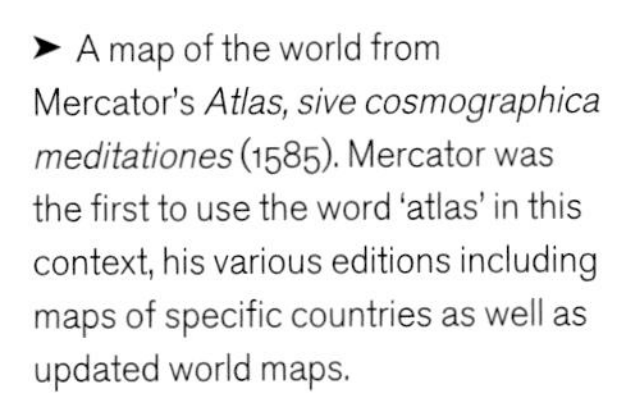

➤ A map of the world from Mercator's *Atlas, sive cosmographica meditationes* (1585). Mercator was the first to use the word 'atlas' in this context, his various editions including maps of specific countries as well as updated world maps.

▲ French sixteenth-century drawing showing a navigator 'shooting a star' to establish his latitude. This early theodolite could measure both vertical and horizontal angles.

angular distance from the equator for each day of the year. Longitude, however, had always been more difficult to establish. The theory was well understood: given a prime meridian as the basis of time measurements, each 15° shift in longitude from the meridian corresponded to a difference of one hour of local time from the meridian. Local time could be established astronomically or with a sundial, but to do so you needed to know simultaneously the time at the meridian. One suggestion was to use the Moon as a kind of nocturnal clock, measuring out the hours as it crosses the sky. But the Moon's apparent motion is highly irregular, and sea voyages so long, that such a method was practical only if the navigator had a table of the Moon's motion stretching for years ahead. It was with this aim that the Greenwich Royal Observatory was founded in 1675. It was not until 1767 that the Astronomer Royal, Nevil Maskelyne, published his *Nautical Almanac*, which included tables of lunar distances for every 3 hours for a whole year. By that time John Harrison's marine chronometer was close to being perfected, and it soon became the preferred method for determining longitude at sea. With an accurate clock on board ship set to the time at the meridian, it was then a matter of establishing local time by solar or stellar observation, and the difference between the two times gave the ship's longitude.

Projections were further complicated by the growing awareness that the Earth was not a perfect sphere, but an oblate spheroid – a sphere whose poles are slightly flattened. Newton's demonstration of the Earth's oblateness in the *Principia* was eventually confirmed experimentally. If the Earth were flattened at the poles, then the length of 1° of latitude should increase as one moves from the equator to the poles, and so should the acceleration due to gravity. Expeditions were organized to measure both effects. In 1735

▼ A portable diptych sundial and compass which closes like a book. Made of ivory it is inscribed with the name of the instrument-maker Paul Reinman and dated Nuremberg 1599. The string gnomon could be adjusted to a limited range of latitudes but the pin-gnomon dials were accurate only at their designated latitude.

the Academy in Paris resolved to send missions to Lapland and Peru to measure any differences between a degree of latitude near the pole and the equator. Christiaan Huygens' classic work on the simple pendulum showed that its period of oscillation was related to the value of the gravitational acceleration, discrepancies being noticed as early as 1672 when a pendulum set to beat seconds in Paris had to be shortened in order to keep the same time in Cayenne. Unfortunately, observational errors tended to lead to incompatible results, some even suggesting that the Earth was a prolate spheroid – that is, elongated rather than flattened at the poles. By 1832 the American astronomer Nathaniel Bowditch had at his disposal 52 measurements taken from around the world, from Lapland to the Cape of Good Hope. To his translation of Laplace's *Traité de Mécanique Céleste* he added his analysis of these results and gave the degree of flattening of the Earth, a quantity known as its oblateness or ellipticity, as 1/297, a value which was to be accepted internationally nearly a hundred years later.

Such deviations from a perfect sphere prompted searches for a form of trigonometry which went beyond plane and spherical, and would comfortably handle general spheroids. The sum of the angles of a triangle on a sphere is greater than 180°, but this excess will vary with location if the surface is instead a general spheroid. Adrien-Marie Legendre (1752–1833) did some elegant work on this in 1799,and found a formula relating the sides of a triangle to their excess over 180°. New projections were then defined using the calculus, where one could define the required distortions by means of formulas. Johann Heinrich Lambert (1728–77) published in 1772 a number of different projections, one of which, the conformal conical projection, is still in use. In this projection the Earth is projected onto a cone which touches the sphere at the 'standard parallel'; the cone can then be opened out to create a plane map.

The tools of the trade were rapidly improving. The astrolabe, inherited from the Greeks and perfected by the Arabs, was a kind of analogue computer. By rotating a disk on which was engraved a projection of the sky and the orbits of various heavenly bodies, one could calculate rising and setting times. Each projection was for a fixed latitude, so an astrolabe often came with a number of discs, each for a different latitude. The astrolabe could also be used to calculate the altitude and azimuth of heavenly bodies, to calculate the time, and to measure astronomical distances. The use of altitude and azimuth as the standard measures was due to the Arabs, the altitude being the angle from the horizon, and the azimuth the angular distance from a meridian. Sundials were also a

common way of telling the time, using either the changes in the Sun's altitude during the day or changes in its azimuth. Most dials needed to be orientated with the aid of a compass, but grew greatly in sophistication, taking into account changes in the Sun's speed of motion across the sky. Universal sundials were also made in the seventeenth century which could be used at any latitude, although they still had to be adjusted for one's latitude. The mariner's astrolabe, a fairly simple construction, was replaced by the quadrant. The quadrants, sextants and related instruments used by navigators, astronomers and surveyors became increasingly accurate with the incorporation of optical instruments and finer scales.

The increasing demand for accuracy in measurements on land, at sea or in the sky brought with it an increase in the amount of computation that had to be performed. Adding extra figures for accuracy meant lengthier calculations. The use of logarithms from the seventeenth century was thus a major practical development. Navigators would have tables of trigonometric functions and logarithms to facilitate calculations, although such tables continued to be plagued with printing errors. Savings in time, if not necessarily accuracy, were further achieved with the invention of the slide rule, which was in wide use by the eighteenth century. By then our view of the world was very different to Ptolemy's – the Earth was now a mere planet, an oblate spheroid in orbit around the Sun. In the second half of the twentieth century we would finally see the Earth from a vantage point way above its surface, when artificial satellites began to map the changing geography of our planet from orbit.

▲ *The Astronomer* by Johannes Vermeer, 1668. With improved telescopes and access to the southern hemisphere, astronomers added new stars to the heavens. Celestial and terrestrial globes were largely used as teaching devices and as ornate decorative symbols of the new knowledge.

chapter fifteen

the quintic

$$a_2x^2 + a_1x + a_0 = 0$$

$$x = \frac{-a_1 \pm \sqrt{a_1^2 - 4a_2a_0}}{2a_2}$$

◄ This general quadratic equation has two solutions obtainable from the well-known text-book formula learnt at school. Formulas to solve the cubic and quartic were finally found in the sixteenth century. But no such formula, an algebraic solution, could be found to the quintic. Some suspected that no such solution existed but it was not until the nineteenth century that this was proved to be the case.

Mathematicians in the sixteenth century came across complex numbers almost by accident (Chapter 11). By the eighteenth century complex numbers were established as an extension of real numbers, but handling them was still leading to the odd lapse, as in Euler's *Vollständige Anleitung zur Algebra* ('Complete Introduction to Algebra', 1770), in which he wrote that $\sqrt{-2} \times \sqrt{-3} = \sqrt{6}$ rather than $-\sqrt{6}$, confusing some later writers on the subject. Even Gauss's masterpiece on number theory, his *Disquisitiones arithmeticae* (1801), avoided the use of so-called 'imaginaries'. For our purposes, the most important part of this work is the first ever proof of the Fundamental Theorem of Algebra. Gauss fully realised just how fundamental this theorem was, giving some further proofs over the years, until in 1849 he recast his first attempt but this time employing complex numbers. In modern terms, it states that for any finite polynomial equation with real or complex coefficients, all its roots are themselves either real or complex numbers. This answered in the negative the long-standing question of whether solutions to higher-order polynomials required the construction of higher-order numbers beyond the complex ones.

One of the thorniest problems in algebra at the time was whether the polynomial of order five, the quintic, was solvable by algebraic methods – that is, in a finite number of algebraic steps. In school we now learn the formula for solving quadratic equations, and similar methods have been known for cubic and quartics since the sixteenth century (Chapter 11). But no method could be found for the quintic. The Fundamental Theorem of Algebra may appear to hold out the prospect of a positive answer, but this theorem merely guarantees that solutions exist, it says nothing about the existence of formulas giving exact solutions (approximate numerical and graphical methods already existed). Enter two mathematical geniuses who both suffered tragic fates.

Niels Henrik Abel (1802–29) was brought up in a large and poor family in a small village in Norway, a country destitute after years of war with England and Sweden. A sympathetic teacher encouraged his private studies, but at the age of eighteen the death of his father brought the family's responsibilities upon his young shoulders and fragile health. In 1824 he published a memoir announcing that the quintic was not solvable by algebraic means, and nor for that matter was any polynomial of higher degree. Abel believed this to be his ticket to an academic career, and sent the paper to Gauss at the University of Göttingen; unfortunately, it seems that Gauss did not get round to separating the pages with a knife (in those days a task for the author or reader, rather than the printer) and so never read it. In 1826 the Norwegian government finally gave Abel the means to travel around Europe. Fearing that little joy would come from calling on Gauss personally, he avoided Göttingen and went to Berlin instead. There he struck up a friendship with August Leopold Crelle (1780–1855), engineer and mathematics consultant to the Prussian Ministry of Instruction, who was about to launch his *Journal für die reine und angewandte Mathematik* ('Journal of Pure and Applied

Mathematics'). Abel's work now found a vehicle for wider distribution, and he contributed greatly to the early volumes of the *Journal*, which had immediately established itself as a prestigious and authoritative publication. He included an expanded version of his proof that the quintic is unsolvable. He then left for Paris, where he became very disconsolate, for he found it difficult to gain the support he needed from the French mathematicians. He approached Augustin-Louis Cauchy (1789–1857), then the leading light in mathematical analysis but a difficult man. As Abel himself put it, 'Cauchy is mad and there is nothing that can be done about him, although, right now, he is the only one who knows how mathematics should be done.' If we can possibly justify the slights suffered at the hands of both Gauss and Cauchy, it may be said that the quintic had achieved a certain notoriety, and had attracted the attention of esteemed mathematicians and cranks in equal measure. Abel returned to Norway, increasingly weakened by tuberculosis. He continued sending material to Crelle, but died in 1829, unaware of his growing reputation. Two days after his death, an offer of a position in Berlin arrived at his house.

Abel had shown that, in general, any polynomial of order greater than four could not be solved by radicals such as square roots, cube roots or higher roots. However, the explicit conditions under which particular cases could be solved and their method of solution was left to Galois. The life of Évariste Galois (1811–32) was short and eventful. An outstandingly gifted mathematician, he had a volatile nature that was stoked up by a whole series of injustices. He was unforgiving to those he considered less talented than himself, yet he detested social injustice by those in authority. He showed no early signs of mathematical ability until he read Legendre's *Éléments de géométrie* (published in 1794, it became the leading textbook for a hundred years). He then swiftly devoured the writings of Lagrange and, later, Abel. His enthusiasm, confidence and impatience had a near disastrous effect upon his relationships with his teachers and examiners. Launching himself into the competitive examinations for the École Polytechnique, the cradle of French mathematics, the unprepared Galois duly failed the exam. His bitterness was temporarily kept in check through meeting a new teacher who recognized his gifts. In March 1829 Galois published his first paper on continued fractions, keeping to himself his most important work. Galois submitted these new discoveries to the Academy of Sciences, which Cauchy promised to present but duly forgot. Worse than that, he then lost the manuscript.

Galois's second failure to enter the École Polytechnique has gone down in mathematical folklore. He was used to handling complex ideas in his head, and was infuriated by the nit-picking of the examiners. Realizing his interview was going badly, he flung the board rubber at the face of one of the examiners. Soon after this his father died, committing suicide as a result of a clerical plot against him; there was virtually a riot at his funeral. In February 1830 he wrote three further papers, submitting them all to the Academy of Sciences in the competition for the Grand Prize in Mathematics. The Secretary, then Joseph Fourier, died before reading them, and they could not be found after his death. Such a catalogue of disappointments would have tested any person, and Galois revolted against the Establishment, which he felt had denied him his dues and murdered his father. He rushed headlong into politics as a staunch republican – not a wise decision in the France of 1830. In one last desperate attempt he sent a memoir to Siméon-Denis Poisson, who replied requesting further proofs.

This was to be the last straw. In 1831 Galois was arrested twice, once for allegedly inciting the assassination of King Louis Philippe, then for his own protection, the authorities fearing a republican revolt! This time he was sentenced to six months' imprisonment on a trumped-up charge of illegally wearing the uniform of the disbanded artillery battalion he had joined. Out on parole, he had an affair which seems to have disgusted him as much as everything else. His letters to his devoted friend Chevalier tell of his disillusionment with life. On 29 May 1832 he accepted a duel for reasons that have never become clear. 'I die the victim of an infamous coquette. It is in a miserable brawl that my life is extinguished,' he writes in a 'Letter to All Republicans'. Galois's most famous work was frantically scribbled down the night before the fateful duel. In the margins are pleas of, 'I have no time; I have no time', as he was forced to leave to others the work of fleshing out the intermediate steps which were not essential to understanding the main results. He needed to get onto paper the essentials of his discoveries – the beginnings of what is now called Galois theory. He ended his will to Chevalier, imploring him to 'ask Jacobi or Gauss publicly to give their opinion, not as to the truth, but as to the importance of these theorems.' In the early morning, Galois went to meet his adversary. It was to be pistols at 25 paces. Galois was wounded and died the following morning in hospital, just twenty years old.

Galois built upon previous work by Lagrange and Cauchy, but developed a more general method while making the crucial breakthrough on the quintic. He looked not so much at the original equation or the graphical interpretation, but rather at the nature of the roots themselves. To simplify things, Galois considered only so-called irreducible quintics, that is, those quintics which could not be factorised into lower-order polynomials (as we have already seen that polynomial equations up to the quartic all have formulas to find their roots). In general an irreducible polynomial with rational coefficients is one which cannot be factorised into simpler polynomials which themselves have all rational coefficients. For example, $(x^5 - 1)$ can be factorised into $(x - 1)(x^4 + x^3 + x^2 + x + 1)$, whereas $(x^5 - 2)$ is irreducible. Galois's aim was to establish the conditions under which all the solutions to a general irreducible polynomial equation can be found in terms of radicals.

The key lay in the discovery that the roots of any irreducible algebraic equation are not independent, but could be expressed in terms of one another. These relationships were formalized into a group of all possible permutations, the so-called symmetry group of the roots – for the quintic this group contains $5! = 5\times4\times3\times2\times1 = 120$ elements. The mathematical machinery of Galois theory is very involved, and may be partially responsible for its initially slow uptake. But in raising the level of abstraction from the algebraic solutions of equations to the algebraic structure of their associated group, Galois was able to predict the solvability of an equation from the properties of such a group. Not only that, but the theory also furnished a method by which the roots themselves could be found. With respect to the quintic, Joseph Liouville (1809–82), who in 1846 published much of Galois's work in his *Journal de Mathématiques Pures et Appliqées*, remarked that Galois had proved the 'beautiful theorem' that, 'In order that an irreducible equation of prime degree be solvable by radicals, it is necessary and sufficient that all its roots be rational functions of any two of them.' As this is not possible for the quintic, the quintic cannot be solved by radicals.

In the space of three years the mathematical world had lost two of its brightest new stars. Recriminations and soul-searching followed, the names of Abel and Galois gaining their rightful recognition only posthumously. In 1829, Karl Jacobi learned of Abel's 'lost' manuscript through Legendre, and in 1830 a diplomatic storm erupted when the Norwegian consul in Paris demanded that Abel's paper be found. Cauchy eventually found the memoir, only for it to be lost again by the Academy's editorial team! In the same year Abel was awarded the Grand Prize for Mathematics jointly with Jacobi – but by then he was already dead. His memoir was eventually published in 1841. In 1846, Liouville edited some of Galois's manuscripts for publication, and in his introduction lamented that the Academy had originally rejected Galois's work because of its obscurity and that, 'Clarity is, indeed, all the more necessary when one essays to lead the reader farther from the beaten path and into wilder territories.' He continues, 'Galois is no more! Let us not indulge in useless criticisms; let us leave the defects there and look at the merits.' The fruits of Galois's brief life amount to no more than sixty-odd pages. The controversy rumbled on. The editor of the mathematical journal for candidates for the École Normale and the École Polytechnique commented on the Galois affair that, 'A candidate of superior intelligence is lost with an examiner of inferior intelligence. Hic ego barbarus sum quia non intelligo illis [because they don't understand *me*, *I* am a barbarian].'

first of all, the second page of this work is not encumbered with names, forenames, qualities, titles, and elegies of some miserly prince whose purse will be opened with the fumes of incense — with the threat of being closed when the censer-bearer is empty. you will not see, in characters three times larger than the text, a respectful homage to someone of high position in the sciences, to a wise protector — something indispensable (i was going to say inevitable) to someone twenty years old who wants to write. I do not say to anyone that I owe to their advice and their encouragement everything that is good in my work. I do not say it because that would be a lie. If I had to address anything to the great in the world or the great in science (at the present time the distinction between these two classes of people is imperceptible) I swear that it would not be in thanks. I owe to the ones that I have published the first of these two memoirs so late, to the others that I have written it all in prison, a place it would be wrong to consider a place of meditation, and where I am often amazed at my self-restraint and keeping my mouth shut in the face of my stupid ill-natured zoiles; I think I may use the word zoiles without fear of being immodest when my adversaries are in my mind so base. It is not my subject to say how and why I was detained in prison, but I must say how my manuscripts have been lost most often in the cartons of Messieurs the members of the Institute, although in truth I cannot imagine such thoughtlessness on the part of men who have the death of Abel on their consciences. For me, who does not want to be compared with that illustrious mathematician, it suffices to say that my memoir on the theory of equations was deposited, in substance, with the Academy of Sciences in February 1830, that extracts from it had been sent in 1829, that no report on it was then issued and it has been impossible to recover the manuscript.

Galois, unpublished preface, 1832

chapter sixteen

new geometries

◀ *Circle Limit IV* by M. C. Escher (1898–1972). This gives an artistic representation of hyperbolic geometry, a two-dimensional non-Euclidean geometry suggested by Felix Klein (1849–1925) as an alternative to Eugenio Beltrami's pseudosphere. In this geometry, the sum of the angles of a triangle is less than 180° and Euclid's parallel postulate does not hold.

Ever since Euclid's *Elements* appeared in the third century BC, Euclidean geometry (Chapter 4) had been heralded as the most perfect of mathematical systems. Founded on the most basic assumptions, it builds an increasingly spectacular architecture of mathematical theorems. Euclidean geometry was the archetypal axiomatic deductive system. However, this temple to geometry had a small blemish, an itch that mathematicians had continued to scratch. Euclid's now infamous fifth postulate states that, 'If a straight line falling on two straight lines makes the interior angles on the same side less than two right angles, the two straight lines, if produced indefinitely, meet on that side on which the angles are less than the two right angles.' Also known as the parallel postulate, it simply says that if two lines are not parallel, they will eventually meet at a point. All agreed that the postulate was true, but it seemed too complicated to warrant it being essentially an axiom, a starting point, of the *Elements*. Initial efforts were therefore directed towards proving that it was actually a theorem which could be proved from other axioms, rather than a postulate. Many deluded themselves they had succeeded in this, but closer scrutiny always showed that new assumptions always crept into their proofs, which were essentially restatements of the fifth postulate. A more obvious replacement was proving hard to find.

Mathematicians continued to investigate the fifth postulate, notably al-Khayyami in the eleventh century and Nasir al-Din al-Tusi in the thirteenth, whose work translated into Latin inspired the Jesuit mathematician Girolamo Saccheri (1667–1733). In his last year Saccheri published a book entitled *Euclides ab omni naevo vindicatus* ('Euclid Cleared of Every Flaw') in which he sought to prove the parallel postulate by disproving as absurd all other possible postulates. He constructed what is now known as the 'Saccheri quadrilateral', with two pairs of 'parallel' lines and three different hypotheses about the sum of the internal angles of the quadrilateral: that the sum was either less than, equal to or greater than four right angles, or 360°. If he could show that the first and third hypotheses resulted in a logical inconsistency, then he would have shown that the middle hypothesis, which is equivalent to the parallel postulate, is the only self-consistent geometry.

> You must not attempt this approach to parallels. I know this way to its very end. I have traversed this bottomless night, which extinguished all light and joy of my life. I entreat you, leave the science of parallels alone ... I thought I would sacrifice myself for the sake of the truth. I was ready to become a martyr who would remove the flaw from geometry and return it purified to mankind. I accomplished monstrous, enourmous labours; my creations are far better than those of others and yet I have not achieved complete satisfaction ... I turned back unconsoled, pitying myself and all mankind.
>
> Letter from Wolfgang to his son, János Bolyai

Saccheri easily dismissed the third hypothesis as leading to logical contradictions. However, the first hypothesis led to no logical problems, indeed he was able to prove theorem after theorem using this new postulate. Saccheri was building before his very eyes the very first non-Euclidean geometry – but he refused to believe it. Remember, his whole aim was to *disprove* the validity of this hypothesis, not to construct a new geometry.

Reverting to his clerical training, he dismissed this new geometry on spurious theological grounds. Future mathematicians were less disbelieving.

This apparent obsession with the fifth postulate had a deeper significance beyond logical cleanliness. The nature of physical space itself was at stake. Euclidean geometry was not only a coherent and robust mathematical system, but it was also the way space itself was structured – the shortest line between two points *was* a straight line, not only in theory but also in practice. On the face of it there was already a well established geometry in which even this was not true, classical spherical geometry. The shortest line between two points on a sphere is a portion, or arc, of a great circle joining those points. Also, the sum of the angles of any triangle on a sphere add up to more than 180°. So what was the fuss all about? It came down to a distinction between what were called the intrinsic and extrinsic properties of a geometry. Extrinsic properties are those that can be deduced from outside the system; intrinsic ones from the inside. For example, the rules for spherical geometry can be deduced from observing a sphere from the outside, like holding a ball in your hand, but how can we tell from a purely geometrical point of view whether we live on a sphere or not? Can we tell *geometrically* whether we live on a flat Earth or a spherical one? Or put another way, are there any *intrinsic* properties which are different for a plane and a sphere? These relatively simple notions are important in considering the true nature of three-dimensional space, where we only have access to intrinsic properties.

I have not yet made the discovery but the path which I have followed is almost certain to lead to my goal, provideed this goal is possible. I do not yet have it but I have found things so magnificent that I was astounded. It would be an eternal pity if these things were lost as you, my dear father, are bound to admit when you see them. All I can say now is that I have created a new and different world out of nothing. All that I have sent you thus far is like a house of cards compared with a tower.

Letter from János to his father, Wolfgang

Johann Heinrich Lambert (1728–77) came within a whisker of a full non-Euclidean system. In his *Theory of Parallel Lines* (1766) he used a similar method to Saccheri's to show that the three scenarios were equivalent to having a triangle whose angles were less than, equal to or greater than 180°. He also demonstrated that spherical geometry is similar to the third of these cases, and speculated that the first case might correspond to a geometry on a sphere of imaginary radius. Replacing a real radius for an imaginary one led to theorems and formulas in what came to be called hyperbolic geometry, in which the familiar sin x and cos x are replaced by sinh x and cosh x (pronounced 'shine x' and 'cosh x'). So, though the idea seems physically implausible, it was mathematically sound. Lambert's speculations were later shown to be not far from the truth.

By the beginning of the nineteenth century, when all attempts at proving the fifth postulate had ended in failure, it was dawning on mathematicians that self-consistent geometries other than Euclid's may indeed be possible. In one of those amazing instances of simultaneous discovery, we find two unknown mathematicians stepping into the limelight.

➤ *Möbius Strip II* by M. C. Escher (1898–1972). The Möbius strip was one of the first exotic topological spaces, with only one surface bounded by only one side. Two Möbius strips 'zipped' together form a Klein bottle (see page 132).

Nicolai Ivanovich Lobachevsky (1793–1856) was the son of a minor official in Russia who died when Nicolai was only seven, leaving a widow and three sons in financial difficulties. Moving to Kazan, the children did well academically, and Lobachevsky shone particularly brightly. He attended the recently founded University of Kazan at the age of fourteen and there came into contact with distinguished professors, many from Germany. At the early age of twenty-one Lobachevsky became a member of the teaching staff and a full professor two years later. As a patient, methodical and hard-working man he earnt the respect of his peers, who rewarded him with a succession of thankless administrative tasks. He found himself university librarian as well as curator of the shambolic university museum. With no assistants, he did all the work himself, bringing some order to both library and museum.

In 1825 the government finally appointed a professional curator to the university, who subsequently used his own political influence to hand Lobachevsky the top job. By 1827 he was rector of the university, and with customary good nature set about reorganizing the staff, liberalizing the teaching and building up the infrastructure, including the founding of an observatory. The university was his life, and he loved it. In 1830, when cholera swept through Kazan, Lobachevsky ordered all students, staff and their families to seek sanctuary within the walls of the university. With strict sanitary rules enforced, only 16 died out of 660 people. In 1846, in recognition of his untiring work for the good of the University of Kazan, the government inexplicably removed him as both rector and professor. His colleagues and friends pleaded with the authorities but to no avail. Lobachevsky's eyesight was now failing badly, but he continued his mathematical researches. His final publication was dictated, as by then he was totally blind.

In 1826 Lobachevsky delivered his first paper to the University (in French, then the common language of scholarship), in which he outlined some of his ideas on geometry, and it took three years for the *Kazan Messenger* to publish *On the Principles of Geometry*. Thus 1829 is the official date for the birth of non-Euclidean geometry in its Lobachevskian form. In this paper he stated that the fifth postulate cannot be proved, and he built a new geometry by replacing the postulate with another. He appreciated what Saccheri and Lambert had seen only dimly, and constructed a geometry that was every bit as solid and logical as Euclid's. Even to Lobachevsky, some of the theorems he derived seemed contrary to common-sense notions of space, and he called his discovery 'imaginary geometry'. But he was under no illusion as to the importance of his work. In 1835–38 his *New Foundations of Geometry* appeared in Russian, and in 1840 *Geometrical Investigations on the Theory of Parallels* came out in German. It was on the strength of this book that Gauss recommended Lobachevsky to the Göttingen Scientific Society, to which he was duly elected in 1842. However, Gauss refused to praise his work in print, thereby contributing to the slow uptake by the mathematical community of these revolutionary ideas. This was a real disappointment to Lobachevsky,

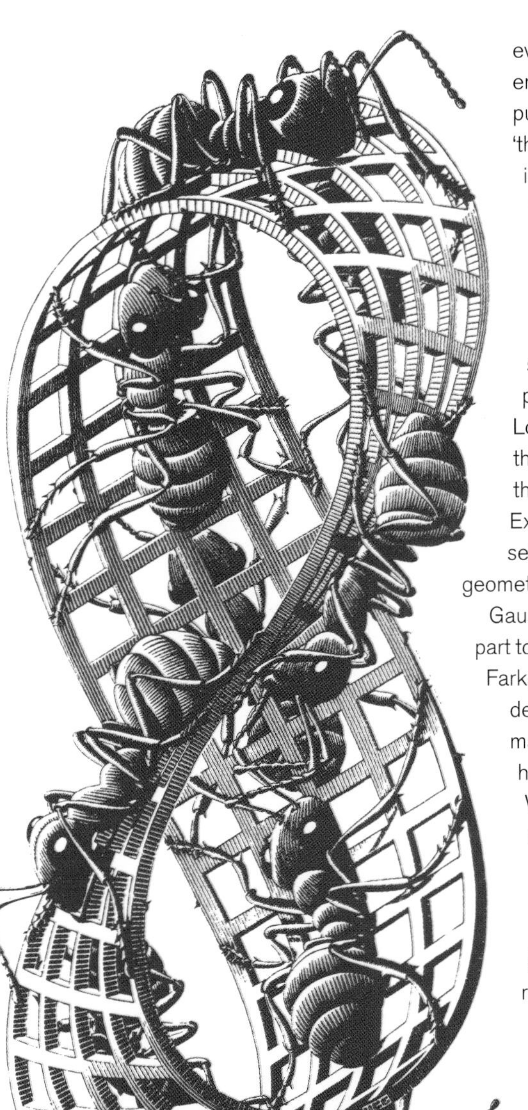

even more so following his dismissal from his university and his encroaching blindness. In 1855 his last book, *Pangeometry*, was published simultaneously in French and Russian. Lobachevsky, 'the Copernicus of Geometry', died the following year. The physical interpretation of non-Euclidean geometry was furnished by Eugenio Beltrami (1835–1900) who showed that the surface of the pseudosphere satisfied Lobachevsky's geometry, and for that matter Lambert's earlier work.

Lobachevsky's new postulate can be explained as follows. Imagine a straight line drawn indefinitely long, and pick a point in space not on this line. Euclid's postulate states that through this point there is one and only one line which is parallel to the first line. Lobachevsky said that more than one line can be drawn through this point, and that all these lines are 'parallel' to the original line in the sense that they do not have a point of intersection with it. Expressing this in mathematical terms leads to an odd but perfectly self-consistent geometry. In fact, there are an infinite number of such geometries each depending on the 'angle of parallelism'.

Gauss's reticence to promote Lobachevsky's work may be attributed in part to a desire on his part to show an even-handedness toward his friend Farkas Bolyai, whose son János Bolyai (1802–60) had simultaneously developed non-Euclidean geometry. Farkas was a provincial mathematics teacher in a backwater of Hungary (now in Romania) who had become engrossed in an attempt to prove the fifth postulate. When this task was taken up by his son, he despaired of the futility of this course of action, and wrote to János saying, 'For God's sake, I beseech you, give it up. Fear it no less than sensual passions because it, too, may take all your time, and deprive you of your health, peace of mind, and happiness in life.' Undaunted, or possibly even encouraged, by such sentiments, János continued his researches, and in 1829 reached virtually the same conclusion as had Lobachevsky.

Bolyai developed what he called the 'absolute science of space' on the same principles as Lobachevsky. His father published the piece as an appendix to a treatise of his own that he was about to publish. The work is dated 1829, the same year as Lobachevsky's article, but it was not in print until 1832. Tucked away at the back of an unfashionable book, it might have been completely lost to history, save for the fact that Farkas was a friend of Gauss, to whom he sent a copy. Gauss's dry response was to express his approval but to refrain from public

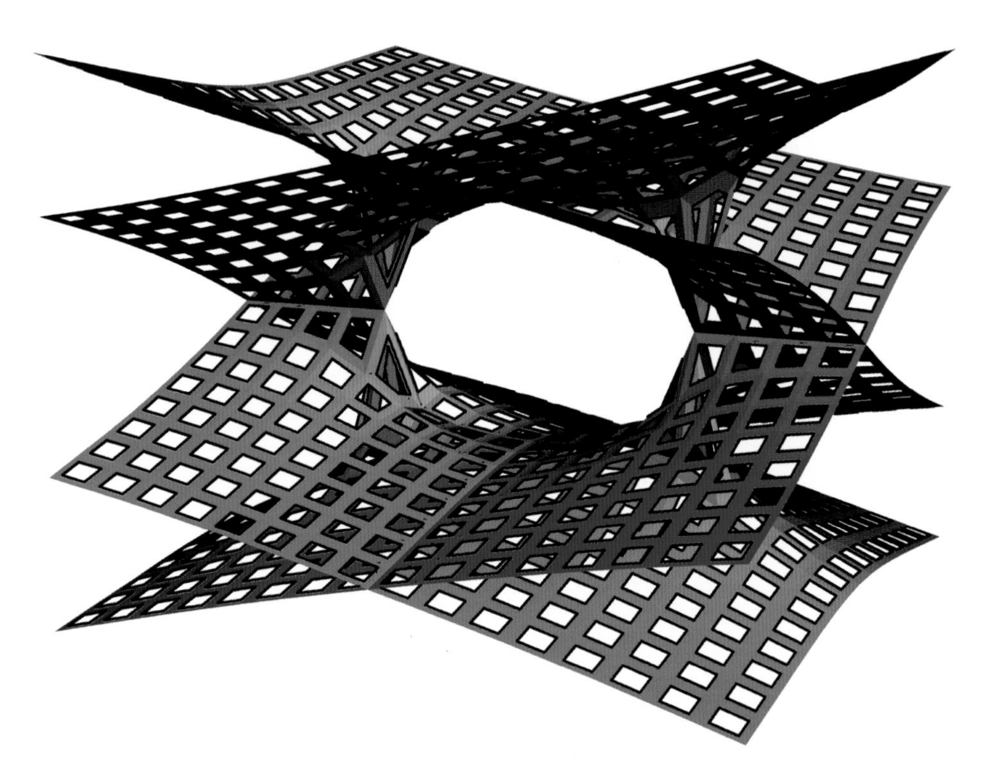

▲ The Riemann surface of the function $(z^2 - 1)^{1/4}$. To get a feel for a Riemann surface, in a two-dimensional complex plane the number $i = \sqrt{-1}$ can be interpreted as a 90° anti-clockwise rotation. Four such rotations of the point (1,0) would return it to where it started, that is, $i^4 = 1$, but Riemann sought to distinguish these two by creating multiple complex planes stacked upon each other and linked like a kind of corkscrew.

support, stating that to praise the work would have been merely to praise himself as he had held the same views years before. János was devastated by the response, fearing that he was going to be robbed of his discoveries. He refused to publish anything else.

Gauss's reluctance to recognize the work of both Lobachevsky and Bolyai seems churlish. Yes, Gauss had certainly thought about these problems, but there is no evidence that he had explored all the implications of non-Euclidean geometry. A helping hand from such an acknowledged master may well have saved Bolyai's career and Lobachevsky's health. Gauss himself had come to the subject from a rather different direction. In looking at lines upon a surface, he came upon the theorem that the 'curvature' of a surface was related to the metric used (that is, to the mathematical expression used for the distance between two points). Gauss showed that curvature was independent of the space in which the surface existed; it was an intrinsic property related to the sum of the angles of a triangle on such a surface. In this context, the similarities with non-Euclidean geometry are obvious.

Having chipped away at the fifth postulate for over two thousand years, the demolition of this pillar of Euclidean geometry made the whole edifice coming crashing down.

▼ Riemann surface of the function $(z^4)^{-1/4}$, where z is a complex number. Bernhard Riemann's 1854 lecture heralded a new and wider perspective on the subject of geometry and he is rightly considered the new Euclid.

Ironically, the discovery of geometries in which Euclid's parallel postulate was not true vindicated Euclid by including it as a necessary postulate despite its apparently complicated form. Euclidean geometry remained logically self-consistent, but it was now merely one of many possible geometries, and hence it was no longer obvious that it was *the* geometry of space itself. The developing awareness of intrinsic properties of space gained in importance as a method of investigating the real geometry of space, as there was no way we could know from the outside! There was a danger that geometry might descend into a menagerie of exotic curiosities, but one mathematician was to give a wholly new definition of what geometry was all about.

Bernhard Riemann (1826–66), a pastor's son, had a modest upbringing but secured a good education in Berlin and Göttingen, where in 1854 he became *Privatdozent*, an assistant professor. To obtain such a position, the University of Göttingen demanded a *Habilitationsshrift*, or doctoral lecture. This was to be the most spectacular doctoral lecture in the history of mathematics. The thesis, entitled 'On the Hypotheses that Lie at the Foundations of Geometry', described in the broadest possible terms what constituted geometry as a subject. This was a long way from the ruler and compass of Euclid. Riemann defined geometry as the study of manifolds – bounded or unbounded spaces of any number (possibly an infinite number) of dimensions, together with a coordinate system and a metric for defining the shortest distance between two points. In Euclidean three-dimensional geometry the metric is given by $ds^2 = dx^2 + dy^2 + dz^2$, the differential equivalent of the Pythagorean theorem. These manifolds are space itself, with no external frame of reference. The curvature of space was thus defined wholly in terms of intrinsic properties of the manifolds in any kind of space. For Riemann, geometry was essentially about sets of ordered *n*-tuples, combined according to certain rules; his ideas on spaces were so general as to be almost *non*-spatial, and any relationship between variables could be considered a 'space'. If no metric is defined for a system then we are in the branch of mathematics known as topology, which deals with how regions of space are connected to one another.

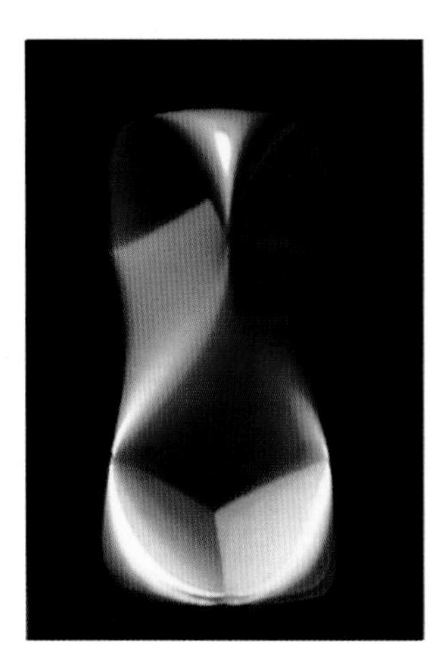

▲ *Etruscan Venus: Red* (1986, NCSA) is a still image of an animated three-dimensional projection of a four-dimensional surface. Topologically equivalent to a Klein bottle this surprising view was so-named because of its obvious similarity to the female figure.

▼ A glass Klein bottle, which has only one surface and no boundary – this is difficult to represent in three dimensions but imagine that one can walk across where the surface cuts through itself.

Riemann had invented tools which are now in the toolkit used by all mathematicians. It is no surprise that on this occasion the normally parsimonious Gauss actually expressed enthusiasm for the work of someone else. Within Riemann's expanded view of geometry, we find that Euclidean geometry is that space defined by a constant curvature of zero; Lobachevskian geometry has a curvature of −1, and spherical geometry a curvature of +1. Although Riemann could be considered the new Euclid, his name is associated with a very specific geometry, that of interpreting the plane as a mapping of the sphere.

Riemann later contributed to theoretical physics and his general study of metric curved spaces ultimately paved the way for general relativity. The space we live in was no longer Euclidean, but we now had the mathematical tools to explore the true geometry of the universe.

In geometry I find certain imperfections which I hold to be the reason why this science, apart from transition into analytics, can as yet make no advance from that state in which it has come to us from Euclid. As belonging to these imperfections, I consider the obscurity in the fundamental concepts of the geometrical magnitudes and in the manner and method of representing the measuring of these magnitudes, and finally the momentous gap in the theory of parallels, to fill which all efforts of mathematicians have been so far in vain.

Nikolai Ivanovich Lobachevsky, *The Theory of Parallels*, 1840

chapter seventeen

dialects of algebra

$$H = a\begin{bmatrix} 1 & 0 \\ 0 & 1 \end{bmatrix} + b\begin{bmatrix} i & 0 \\ 0 & -i \end{bmatrix} + c\begin{bmatrix} 0 & 1 \\ -1 & 0 \end{bmatrix} + d\begin{bmatrix} 0 & i \\ i & 0 \end{bmatrix}$$

◄ This is the matrix representation of Hamiltonian's quaternions, which proved to be important in the theories of electromagnetism and quantum mechanics.

We saw in Chapter 11 how algebra was liberated from the dimensional shackles of geometry, and how, from Descartes onwards, the symbols of algebra – those *x*'s and y's – could stand for any numerical value and be combined in any manner consistent with the rules of arithmetic. In this chapter we shall look at developments in algebra brought about by Britain embracing and then extending methods developed on the Continent. The resulting proliferation of different dialects of algebra led to a fundamental re-evaluation of what mathematics itself was really about.

THE MAIN RULES OF THE ALGEBRA OF ARITHMETIC
FOR ANY NUMBERS X, Y AND Z

$x+y=y+x$	addition is commutative – the sum of two numbers is independent of the order in which they are added
$x.y=y.x$	multiplication is commutative
$x+0=x$	addition has an identity, zero, which leaves every number unchanged
$x.1=x$	multiplication has an identity, one, which leaves every number unchanged
$x.(y+z)=x.y+x.z$	multiplication is distributive over addition

British analysis had lagged behind the rest of Europe. Much of the blame was laid at the feet of the Newtonian fluxional notation and its inferiority to the Leibnizian symbolism, d*y*/d*x*. The reorientation by the British, although it was met with initial reluctance, yielded some highly influential breakthroughs. In 1817, when George Peacock (1791–1858) was appointed an examiner of the mathematical tripos in Cambridge, differential notation finally replaced fluxional symbols. In the words of Charles Babbage, the aim of the Analytical Society, founded in 1813, was to promote 'the principles of pure d-ism as opposed to the dot-age of the university'; another aim of the Society was 'to leave the world wiser than we found it'. Peacock set out in his *Treatise on Algebra* (1830) to establish algebra as a 'demonstrative science'. The first step was to separate arithmetical algebra from symbolic algebra: the elements of arithmetic algebra were numbers and arithmetic operations, whereas symbolic algebra is 'a science, which regards the combinations of signs and symbols *only* according to determinate laws, which are altogether independent of the specific values of the symbols themselves'. This apparently vague statement opened the door to investigations of algebra in general.

Through sheer determination and brilliance, a previously unknown elementary school teacher from Lincoln, George Boole (1815–64), was to write what is now considered the first work on algebraic logic. Boole had become friends with Augustus De Morgan, whom he supported in a controversy over logic with the Scottish philosopher Sir William Hamilton (1788–1856) – no relation to the Irishman Sir William Rowan Hamilton. The dispute is now of no importance, but it did inspire Boole, a self-taught mathematician and linguist, to publish in 1847 a short work entitled *The Mathematical Analysis of Logic*. The same year saw the publication of De Morgan's own *Formal Logic*. Two years later, most probably with the support of De Morgan, Boole was appointed professor of mathematics at the newly established Queens College in Cork. Boole was adamant that logic should be seen as a part of mathematics rather than metaphysics, and that the rules of logic were not to be derived from ordinary language but were to be constructed from purely formal elements. Only when the logical structure was in place could a linguistic interpretation be placed upon it. He rejected the view that mathematics was the science of numbers and magnitudes, a view going all the way back to the Greeks, but held that any consistent symbolic logical system was a part of mathematics. For the first time, we have the clearly stated view that mathematics is not so much about content but about structure. Boole's *Investigation of the Laws of Thought* (1854) clarified these ideas and established both formal logic and a new algebra, now referred to as algebraic logic. Boolean algebra is essentially an algebra of classes of things, and the variables such as x no longer denote numbers but rather the mental act of choosing a class from a given universe. For example, x could be the class of 'men' from a universe of 'humans'. The symbols follow the same rules as in arithmetic algebra, except for the additional axiom that $x^2 = x$. In arithmetic this is an equation which is true only when x is equal to 1 or 0, but in Boolean algebra it is always true – choosing the class of 'men' twice is exactly the same as choosing it once. Also, Boole assigned the symbols 1 and 0 specific meanings, 1 being the 'Universe' and 0 being 'Nothing'. These ideas are now at the heart of the world's computing revolution, and we shall look at them again in more detail in Chapter 23.

Augustus De Morgan (1806–71) was a staunch supporter of the new algebra. Born in India, he attended Trinity College, Cambridge, but was not made a Fellow at either Oxford or Cambridge because, although a member of the Church of England, he refused to submit to the theological examinations necessary to obtain his MA. Instead, at the age of twenty-two, he was made professor of the newly established and secular London University, later University College, London. He pushed the ideas of Peacock to their limits, and as early as 1830 stated that, 'With one exception, no word or sign of arithmetic or algebra has one atom of meaning throughout this chapter, the object of which is symbols and their laws of combination, giving a symbolic algebra which may hereafter become the grammar of a hundred distinct significant algebras.' His one exception was the symbol for equality, in that in an expression $x = y$ the symbols x and y must have the same meaning. This may sound strange coming from a book entitled *Trigonometry and Double Algebra* (1830), the 'double algebra' referring to the binary nature of complex numbers as opposed to the 'single algebra' of real numbers. But De Morgan did not appear to fully grasp the extent of his own pronouncement: having seen the similarity between the single and double algebras, he believed that a triple or quadruple algebra was not possible. In this he was proved to be much mistaken.

Despite both his parents having died while he was a young boy, William Rowan Hamilton's gifts shone through early. A gifted linguist, he was reading Greek, Hebrew and Latin by the age of five. He entered Trinity College, Dublin, and while still an undergraduate at the age of twenty-two, he was appointed Royal Astronomer of Ireland, Director of the Dunsink Observatory and Professor of Astronomy. One of his favourite themes was that space and time were indissolubly connected, with geometry being the science of space and algebra the science of time. In 1833 he presented to the Royal Irish Academy the definitive view of complex numbers $a + ib$ as ordered pairs (a, b) with the now standard geometric interpretations for addition and multiplication:

$$(a, b) + (c, d) = (a + c, b + d)$$

$$(a, b).(c, d) = (ac - bd, ad + bc)$$

He then tried to extend the system of two-dimensional complex numbers to three dimensions. On the surface this seemed fairly simple – just define $z = a + ib + jc$, with length equal to $\sqrt{(a^2 + b^2 + c^2)}$. Defining addition was easy enough but multiplication just would not work: it would not commute. This and higher-dimensional numbers taxed him for ten years. Then, on 16 October 1843, he was walking with his wife along the Royal Canal and had a flash of inspiration: use quadruples, rather than triples, and abandon the commutative law. So the quadruple is $z = a + ib + jc + kd$ with $i^2 = j^2 = k^2 = ijk = -1$. This meant that $ij = k$ but $ji = -k$, so commutativity was lost. But the whole structure was self-consistent and a new algebra was created. Hamilton stopped in his tracks and carved the formula with a knife on a stone of Broughton Bridge. That day he informed the Royal Irish Academy that he wished to read a paper on quaternions, as he called his quadruples, at the next session.

The importance of this was not only in the creation of a new algebra but in the freedom it gave to mathematics to build further algebras. It was also the first detailed theory of what are now known as non-commutative algebras. The non-commutative property meant that in three dimensions a general sequence of two rotations will give different results depending on which order they are taken in, unlike what happens in two dimensions. Hamilton spent the rest of his life developing his new algebra, publishing *Lectures on Quaternions* in 1853. Much of this work was devoted to applications of quaternions to geometry, differential geometry and physics. As we shall see in the next chapter, James Clerk Maxwell formulated his equations of electromagnetism in quaternion notation. Hamilton believed strongly, almost obsessively, that quaternions held the key to a complete description of the laws of the universe. He died in 1865 before completing his *Elements of Quaternions*, which was edited and published by his son posthumously.

Not only was algebra released from the bonds of geometry in this period, but also geometry was unshackled from spatial concepts (Chapter 16). Both algebra and geometry were increasingly treated as purely abstract constructs, of which the familiar arithmetic algebra on the one hand, and two- and three-dimensional geometry on the other, were merely special cases.

It is in the area of new algebras that we find American mathematics slowly emerging from the birth-pangs of a new nation. Benjamin Peirce (1809–80), professor of mathematics at Harvard and Director of the Geodetic Survey, was very much influenced by Hamilton's work and promoted it widely in the USA. Peirce set about constructing tables for 162 different algebras. Each algebra started with two to six elements, which could be combined under two operations, with multiplication associative over addition. An additive identity 0 was assumed for each algebra, but not necessarily a multiplicative identity 1. Each of these 'linear associative algebras' was laid out in an array. One gets an idea of the economic difficulties in the States from the fact that here was a Harvard professor in the 1870s forced to publish his work as a lithograph written by a lady scribe, with only 100 copies ever printed. Benjamin's son, Charles Sanders Peirce (1839–1914), carried on his father's work and showed that, of all the 162 algebras, only three had the operation of division defined uniquely: arithmetical algebra, complex number algebra and the algebra of quaternions. Back in England, William Kingdon Clifford (1845–79) developed what are now known as Clifford algebras, especially the octonians and biquaternions, primarily to study motion in non-Euclidean space. All this is a far cry from the single algebra at the start of the century.

The story here fragments into many interweaving strands. Followers of Boole applied mathematics *to* logic, forming an algebraic logic,

PROPOSITION I

All the operations of Language, as an instrument of reasoning, may be conducted by a system of signs composed of the following elements, viz.:

1st. Literal symbols as x, y, etc., representing as subjects of our conceptions.

2nd. signs of operation, as +, -, x, standing for those operations of the mind by which the conceptions of things are combined or resolved so as to form new conceptions involving the same elements.

3rd. The sign of identity, =.

And these symbols of Logic are in their use subject to definite laws, partly agreeing with and partly differing from the laws of the corresponding symbols in the science of Algebra.

George Boole, *An Investigation of the Laws of Thought*, 1854

whereas Giuseppe Peano expresses mathematics in terms of logic, and later, Bertrand Russell, sought to obtain mathematics *from* logic, an enterprise which can be named logicism. Others, alarmed at the appearance of so many new mathematical structures, began to search for the solid foundations of mathematics – something that would underpin the whole edifice. The practical implications of this search will be looked at in Chapter 23.

Among the minor, yet striking characteristics of mathematics, may be mentioned the fleshless and skeletal build of its propositions; the peculiar difficulty, complication, and stress of its reasonings; the perfect exactitude of its results; their broad universality; their practical infallibility. It is easy to speak with precision upon a general theme. Only, one must commonly surrender all ambition to be certain. It is equally easy to be certain. One has only to be sufficiently vague. It is not so difficult to be pretty precise and fairly certain at once about a very narrow subject. But to reunite, like mathematics, perfect exactitude and practical infallibility with unrestricted universality, is remarkable. But it is not hard to see that all these characters of mathematics are inevitable consequences of its being the study of hypothetical truth.

Charles Sanders Peirce (1839-1914), *The Essence of Mathematics, c.* 1870

chapter eighteen

fields of action

◄ Electromagnetic fields of force illustrating the mutual repulsion between two electrodes of different size but the same polarity, from James Clerk Maxwell's *Treatise on Electricity and Magnetism* (1873).

From the middle of the eighteenth century, developments in calculus went hand in hand with the growth in the mathematical analysis of physical phenomena, particularly motion. Topics included thermodynamics, celestial mechanics, hydrodynamics, and investigations of light, electricity and magnetism. These were all tackled by setting up differential equations to describe the phenomena, and developing the necessary methods of solving these equations. The difficulties in finding unique solutions led to a focus on methods of approximation. Although the phenomena listed above appeared to be physically different, they all related in some sense to the medium of space. From the time of Newton's *Principia* in particular, arguments had raged over the reality of 'action at a distance' – how could gravitation, for example, operate across space? Were gravitation and magnetism different aspects of the same force, or were they different phenomena? Perhaps space was filled by some kind of medium, known as the aether. If so, what was the aether and what were its properties? To illustrate these multiple strands, I shall focus on the history of potential theory and its relationship with electromagnetism.

The Leibnizian calculus was being expanded to cope with more than one independent variable, so that, for example, as well as curves $y = f(x)$ in the plane, curves $z = f(x, y)$ in space could be studied. This was made possible by the introduction of partial differential equations, in which each variable could be differentiated independently of the others. The interactions of particles in motion could be represented by differential equations whose solutions would hopefully yield the paths travelled by the same. The initial Newtonian solutions which gave the elliptical orbits of the planets had been obtained only by making some fairly crude simplifications, such as assuming that the Sun and planets were point masses, and that each planet could be treated independently from all the others. Now that the initial objections to the heliocentric model and non-circular orbits had been finally overcome, work could start on making the model more accurate and sophisticated. One technique was to look at the energy changes within a dynamical system, potential theory being a mathematical way of expressing the physical idea of conservation of energy.

A major concern arose in celestial mechanics when it was discovered that the planets did not follow perfectly elliptical orbits after all, but instead wobbled in their paths. In fact, with increasingly accurate data it was becoming more and more obvious that everything in the Solar System deviated from smooth motion, and this led to the development of perturbation theory, in which a planet's orbital path was considered as resulting not only from the interaction between that planet and the Sun, but also between that planet and all the other planets. This made mathematical analysis extraordinarily difficult, as there were now so many variables to consider. The three-body problem was discussed a great deal: taking a simplified system of just the Sun, Earth and Moon there were no exact solutions. But then in 1747, Euler developed a new technique whereby the

➤ Thomas Wright, *An Original Theory or new hypothesis of the Universe, founded upon the Laws of Nature, and solving by mathematical principles the general phenomena of the visible creation; and particularly the Via Lactea*, 1710. This image represents the idea of an infinity of universes, each centred upon an Eye of Providence.

distances between planets at any instant could be approximated using trigonometric series expansions.

Leonhard Euler (1707–83) is the most prolific mathematician in history. At the University of Basel he was given some private assistance by Johann Bernoulli. (The Bernoulli family produced many distinguished mathematicians over several generations, and are truly a mathematics dynasty.) By 1727 Euler had joined the St Petersburg Academy of Science, recently founded by Catherine the Great. In 1733 Daniel Bernoulli, Johann's son, returned home to Basel having vacated the senior chair of mathematics at St Petersburg to the young Euler. A year later Euler married; he would father thirteen children, although only five survived their infancy. He was later to write that he had made some of his best discoveries holding a baby in his arms surrounded by playing children. But his eyesight was failing him badly. By 1740, he wrote, he had lost the sight of one eye, and by 1771 he was blind in both eyes. In 1741 he accepted an invitation from Frederick the Great to go to Berlin, and a few years later was made the first director of mathematics of the newly founded Berlin Academy of Sciences. Euler returned to St Petersburg in 1766, and despite his encroaching blindness he produced almost half his total work after this date, greatly helped by devoted assistants and a phenomenal memory.

Euler's mathematical output spanned virtually every field of mathematics, including very practical work in cartography, shipbuilding, calendars and finance. But his chief claim to fame is in laying the foundations of mathematical analysis and of analytical mechanics, with such ground-breaking works as the *Introductio in analysin infinitorum* (1748), his *Theory of the Motions of Rigid Bodies* (1765) and major works on the differential and integral calculus. The whole language of functions is his, as is the notation $f(x)$ and a host of now common mathematical symbols such as π for the ratio of a circle's circumference to its diameter, e for the base of natural logarithms, i for $\sqrt{-1}$ and Σ for summation. He saw number theory, geometry and analysis as all supporting one another in the modelling of natural phenomena.

Perturbation theory was leading to more accurate results for planetary orbits, but it led also to the theoretically disturbing conclusion that there was no reason why the planets should remain in their present paths. Small wobbles could easily become magnified to the point where a planet could wander completely out of its orbit – it was almost as if angelic beings were still needed to keep the planets on their course. (In the twentieth century it was found that the dynamics of the Solar System could be explained by chaos theory – see Chapter 24.) The equations required to describe planetary motion increased in number and complexity. In France, analytical methods were preferred to geometrical ones, and this led to ever more unwieldy equations. The analytical approach was characterized by Joseph Louis Lagrange (1736–1813), who developed a system of equations known as the 'Lagrangian', and whose *Méchanique analitique* (1788) had not a single diagram in its 500-odd pages. In 1799 Laplace published the first volume of the encyclopaedic *Traité de mécanique céleste*, which laid particular emphasis on potential theory and perturbations.

Many of the developments in France took place at a time when few mathematicians could escape the political upheavals of the French Revolution. The young Augustin-Louis Cauchy (1789–1857) avoided the worst excesses of the French Revolution as his family took temporary leave of Paris. After graduating from the École Polytechnique he worked on the port facilities for Napoleon's planned invasion of England. But he really wanted to devote himself to mathematics, and after numerous disappointments he was finally given the position of assistant professor of analysis at the École Polytechnique.

Cauchy's output was prodigious: the landmarks are his *Cours d'analyse* (1821) and the *Leçons sur le calcul differential* (1829), while his collected works fill some 27 hefty tomes. But in the political climate of early nineteenth-century France his staunch Catholic views did not go down well in many quarters, and his relationships with his colleagues were often strained. For supporting the Jesuits against the Académie des Sciences and refusing to swear his allegiance to the new regime of 1830, Cauchy was stripped of all his positions, and he went into exile with Charles X. On his return to Paris he was twice passed over for the chair in mathematics at the Collège de France, despite being by far the best candidate. It was only in 1848 that he regained his university positions, on the overthrow of Louis Philippe. Between 1840 and 1847 Cauchy published his four-volume *Exercises d'analyse et de physique mathématique*. Cauchy helped to lay the foundations of real and complex analysis, which form the basis of mathematical physics.

The French approach of approximating functions through truncated power series, and hoping that better approximations could be gained by taking more terms, came in for criticism from many who sought more manageable methods. For example, as late as the 1860s Charles Delaunay published a monster equation taking up a whole chapter, followed by nearly sixty methods for estimating terms. In 1834, William Rowan Hamilton sent a paper to the Royal Society in which he presented the function we now refer to as the Hamiltonian. In one characteristic equation he could describe the motion of any number of point masses moving within a potential. Not only that, but as Hamilton himself explained, his expression also yielded a method of solution, unlike the Lagrangian, which often defeated attempts at solution. From the mid-nineteenth century the methods and language of potential theory were transformed by Riemann's work on geometry (Chapter 16). The new area, which became known as differential geometry, extended the concepts of the calculus to three-dimensional space. Geometric objects such as points, curves and surfaces were described in terms of vectors, and dynamical concepts such as velocity, acceleration and energy could be described by functions and operators acting upon them. In three dimensions there are three different vector operators defined: a gradient operator (known as grad) which takes a scalar function $f(x,y,z)$ to a vector; a rotation operator (curl) which takes a vector to another vector; and a scalar operator (div) which takes a vector to a scalar function. Indeed, as each variable within a dynamical system could be treated as a 'dimension' of the system, Riemann's work on higher-dimensional spaces made differential geometry the perfect vehicle for modelling physical systems within a single framework. It was in the notation of differential geometry that Maxwell expressed his theory of electromagnetism.

By the mid-nineteenth century there was a large body of experimental and theoretical results on electricity and magnetism. In the 1780s Charles Coulomb had discovered by experiment that the electrostatic force between two charged particles followed the

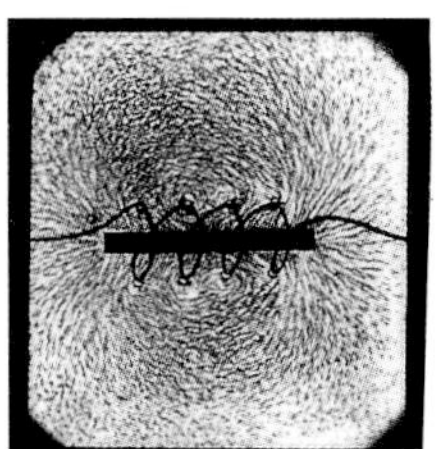
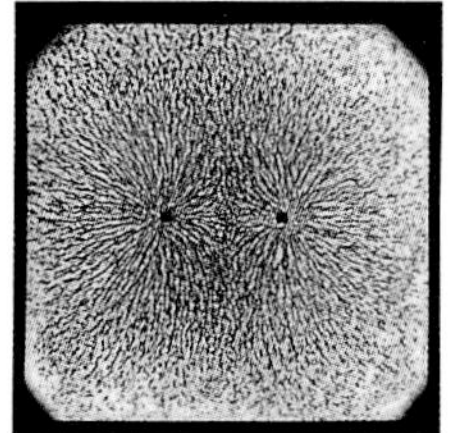
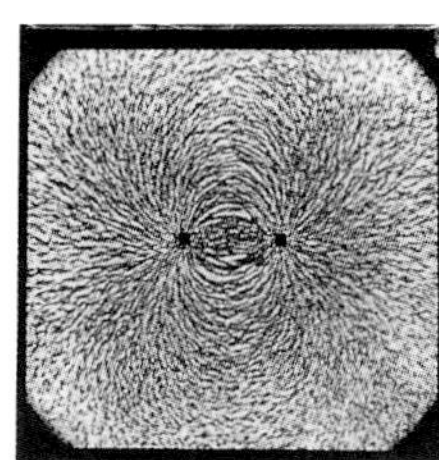
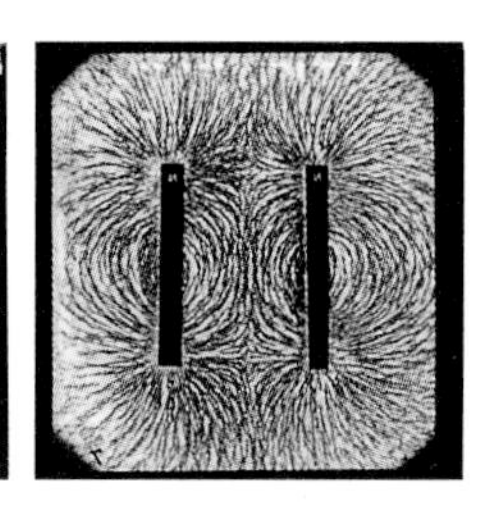
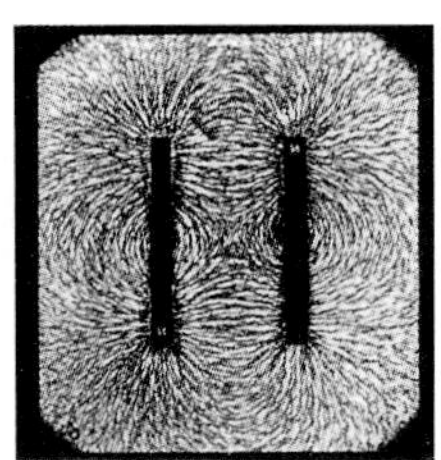

inverse-square law. Scientists could now apply to electrostatic phenomena some of the mathematical models and techniques that had been developed in treating gravitational forces. In 1812, Siméon-Denis Poisson treated electrostatics in a manner reminiscent of Laplace's *Méchanique céleste* of the previous decade. He assumed that electricity consisted of two fluids of opposite charge which existed in all bodies, where like particles repel one another and unlike particles attract. A year later, he derived a partial differential equation that relates the potential to a charge density, now known as the 'Poisson equation'. In 1820, Hans Christian Oersted discovered electromagnetism by showing that a wire carrying a current could cause a magnetic needle to oscillate. This inspired André-Marie Ampère to study the interaction between electricity and magnetism, for which he coined the term 'electrodynamics'. He showed mathematically that the electromagnetic force obeyed an inverse-square law, just as did the electrostatic force. Michael Faraday's discovery of electromagnetic induction showed that electricity and magnetism were inextricably linked. But the physical theories of the time were not up to explaining the phenomena adequately. Ampère's idea of tiny electrical vortices in the aether as a mechanism for the transmission of magnetism, for example, ran up against problems similar to those encountered by Descartes's vortex model of planetary motion.

By analyzing the gravitational interaction between the Earth and the Moon, it became obvious to astronomers that, because of the sizes of the two bodies and the distance between them, they could no longer be taken as point masses: it was now necessary to look at the effect of the whole body. From a point on the Earth, the Moon's gravitational effect is related to both its volume, or mass, and its shape. This relationship between the forces within a body and on its surface were handled mathematically as a relationship between the volume integral and the surface integral. This relationship was codified in 1828 in Green's theorem, named after George Green, who studied mathematics at Cambridge as a mature student, and was later offered a fellowship. This theorem, which Green developed for electromagnetic potentials, could also be used for gravitational potentials.

In 1873 Maxwell published his *Treatise on Electricity and Magnetism*, in which, following Faraday, his key notions are the electric and magnetic fields. Maxwell tried to

◄ Magnetic lines of force as photographed by Sylvanus Thompson in 1878. The force fields are created by current-carrying wires going across the plane of vision or into the page.

▼ Photograph of a positive electrical discharge, made by Alan Archibald Campbell Swinton in 1892.

avoid getting his theories caught up in arguments about the aether and the true nature of space by taking essentially a top-down approach. The theory avoids reliance on microscopic ideas such as charge or current, then not clearly understood, but rather takes a macroscopic approach, assuming the existence of fields which interact with one another and with the medium they travel through. For Maxwell, space was an elastic continuum, and thus able to transmit motion from point to point. Because of this elasticity the medium itself could store kinetic and potential energy. He made copious use of potential theory and differential geometry, originally writing his equations in Hamilton's quaternion notation as well as the Cartesian equivalent. It was Oliver Heaviside who cast Maxwell's equations in the vector forms that are used today.

Maxwell's theories and presentation were not an immediate success. J. J. Thomson charged Maxwell with 'mysticism' regarding his field theories, redolent of the reception Newton received when he proposed gravitation. This period saw confusion reign over the nature of space, and many physicists adapted Maxwell's equations to satisfy their preferred theories. In 1861 Maxwell calculated that the velocity of electromagnetic waves was very close to that of light, inspiring him to incorporate light as a part of the electromagnetic spectrum. In 1888 Heinrich Hertz vindicated Maxwell's theory by experimentally

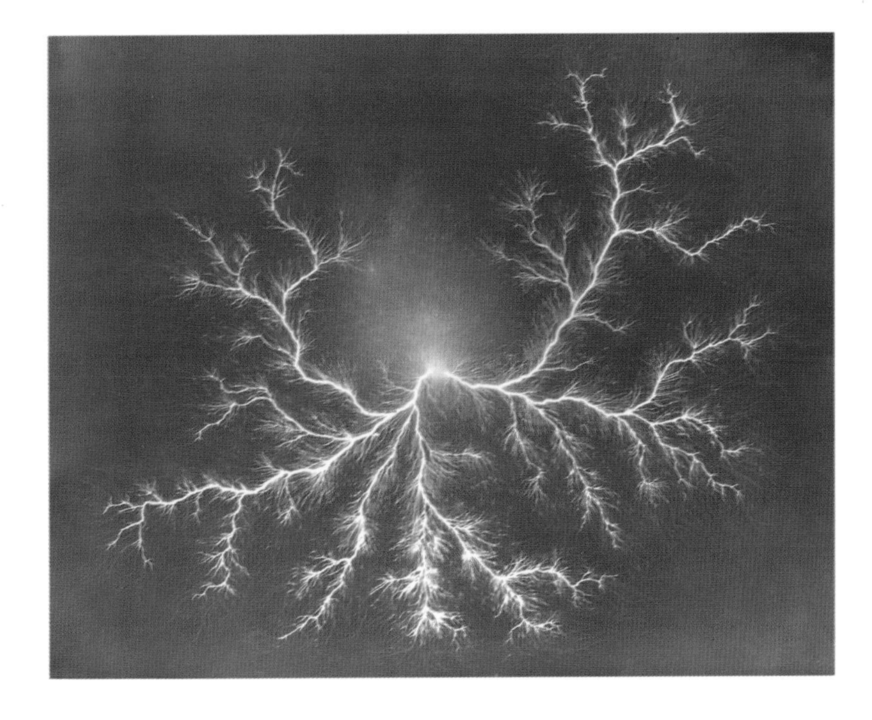

▲ Photograph of a positive electrical discharge, made by Alan Archibald Campbell Swinton in 1892.

demonstrating the existence of electromagnetic waves. At the same time, the experiments of Albert Michelson and Edward Morley showed that if there were an aether then it remained unaffected by motion through it, be it a planet or a beam of light. The old objections to action at a distance vanished in the face of experimental evidence. A major rethink on the whole notion of space and time came in 1905 by Albert Einstein.

Two early uses of Maxwell's equations were in telegraphy and radio communications. Heaviside transformed the subject with his telegraphy equation, which took into account self-inductance in transmission lines that had been overlooked by others. This prompted the introduction of induction coils to boost the signal level along cables such as the transatlantic cable. In 1902 Guglielmo Marconi succeeded in transmitting radio signals across the Atlantic. This presented mathematical physicists with the problem of accurately modelling how electromagnetic waves travelled around the Earth's atmosphere, especially when the receiver lay beyond the visual horizon of the transmitter. The telecommunications industry has never looked back from these pioneering days.

chapter nineteen

catching infinity

◄ In 1890, the Italian mathematician and logician Giuseppe Peano (1858–1932) upset his contemporaries by proposing a curve which filled the whole plane, that is, a one-dimensional object which covered a two-dimensional surface. This is an example of such a curve.

Throughout history, mathematicians and philosophers have struggled with the concept of infinity. The Hellenic dread of actual infinities and their converse, infinitesimals, has resurfaced time and time again, most notably in the definitions of the calculus. The nineteenth century finally confronted the issue head on. Many branches of mathematics have developed through the accumulated work of many minds, but the battle with the infinite and the resulting foundation of set theory were the work of one man, Georg Cantor. The spur was the increasing use of infinite series and the doubts over their validity.

It was Cauchy who rephrased the basic concepts of the calculus in terms of arithmetic rather than geometry (referred to as the arithmetization of calculus). In a reversal of the Greek tradition of giving geometry pride of place as the most rigorous of subjects, the nineteenth-century aim was to recast analysis in the image of arithmetic. This came about largely through the increasing use of functions with multiple variables and complex variables, whose visualization was often impossible.

In 1822 Joseph Fourier (1768–1830) published his classic *Théorie analytique de la chaleur*. Analyzing the flow of heat, Fourier solved the resulting differential equation by what have come to be known as Fourier series. According to Fourier, any function can be represented by an infinite series of sines and cosines, not only smooth continuous functions but even ones with discontinuities or gaps. However, some began to doubt that this infinite series always converged to the required function, and Lejeune Dirichlet proved that it only did so under certain restrictions. The whole concept of a function was further generalized by Dirichlet, who stated that *any* rule relating x and y was a function – it was not necessary to have an analytic expression or equation. As an example, Dirichlet constructed a 'wild function', defined as $y = a$ if x is rational, $y = b$ if x is irrational. This function, which mathematicians would now describe as 'pathological', was discontinuous at every point and therefore not differentiable anywhere, but discussions focused on whether it could be integrated. A resolution of this problem demanded a definition of what exactly was an irrational number.

Galileo mentioned in his analysis of acceleration that taking the infinite series of natural numbers, 1, 2, 3 ... and squaring each number yields the series 1, 4, 9 ... Now, every number in the second series can be put into a one-to-one relationship (correspondence) with a number in the first series, so the two series will have the same number of terms. But the second series has some numbers missing, so it should have fewer terms than the first. Either the two infinities were the same or it was possible to have different kinds of infinity.

Bernhard Bolzano (1781–1848), a priest living in Prague, developed interesting ideas which unfortunately lay undiscovered until much later. He worked on a very similar arithmetization of calculus as Cauchy, who during his exile in Prague met Bolzano once. In *Paradoxien des Unendlichen*, published posthumously in 1850, Bolzano showed that paradoxes such as Galileo discovered are common not only among natural numbers but

▼ David Hilbert (1862–1943) suggested a similar space-filling curve to Peano: in this case, a one-dimensional line which fills a three-dimensional cube. Such counter-intuitive ideas drove mathematicians to examine more closely the nature of numbers, ideas of space and the vague concept of infinity.

also among real numbers. For example, a line segment of length one unit has the same number of reals as a line twice its length, which is very counter-intuitive. This Bohemian philosopher seems to have come very close to recognizing that the infinity of reals is of a different type to the infinity of the naturals. He also made his own contribution to the growing list of pathological functions which succeeded in breaking the established rules of calculus.

This dual concern with properties of functions and of numbers was not accidental. If any function could be expressed as an infinite series, say a Fourier series, then it was important to verify that the series converged to the function for every value of x, so-called point-wise convergence. As it is tedious to check this for each series, various convergence criteria were suggested, all of which necessitated very clear thinking about the idea of an infinite sequence of numbers approaching a given number. Cauchy, with his Hellenistic abhorrence of actual infinities, uncharacteristically slipped into a circular argument, in one place defining an irrational number as the limit of a sequence of rationals, in another defining the converse. Karl Weierstrass attempted to rid the reliance of irrationals on limits and sought to define them not as the limit of a sequence but as the sequence itself.

Meanwhile, Bernhard Riemann reformulated the concept of an integral into what is today still taught in undergraduate classes. The Dirichlet function mentioned above still did not have an integral in Riemann's definition. Adding to the sport of finding unpalatable functions, Riemann found one of his own, which is discontinuous at an infinity of points but for which the integral not only exists but defines a continuous function, which in turn, however, fails to have a derivative for that same infinity of points. The fundamental theorem of calculus was further called into question.

What was still needed was a real understanding of what an irrational number was, and hence a clear definition of a real number. By the 1850s it was accepted that real numbers could be divided into two types in two different ways: as rationals and irrationals; and as algebraic and transcendental. Rationals were all numbers of the form m/n, that is to say, any positive or negative fraction including whole numbers and zero. Irrationals were numbers that are not rational, such as $\sqrt{2}$ or π. Algebraic numbers were those numbers that are solutions of finite polynomial equations with integer coefficients, thus including numbers such as $\sqrt{2}$ but not π. The transcendentals were numbers which

▲ These reflective spheres are an example of self-similarity, the central sphere sprouts further spheres of half the radius, each of which gives birth to further spheres. If we continue this process the surface area tends towards infinity while the total volume remains bounded and finite.

were not algebraic. We can immediately see that both irrationals and transcendentals are merely defined by what they are *not*, and it was not clear whether they had any particular properties of their own. The year 1872 saw the publication of key works on this topic by Richard Dedekind (1831–1916) and Georg Cantor (1845–1918), who that same year had struck up a close friendship. Both were at relatively minor institutions – Dedekind at the *Polytechnikum* in his home town of Brunswick, Cantor at the University of Halle – but their work was to have an enormous impact in the mathematical world.

Given a continuous real number line what, asked Dedekind, was the difference between a rational and an irrational number. Leibniz, for example, thought that the 'continuousness' of points on a line was related to their density; that is, for any two points there is always a third between them. However, the rationals also have this property but are not continuous. Rather than continuing to look at ways of gluing together points to form a continuum, Dedekind took the opposite view and sought to establish continuity in terms of defining cuts in the line segment. Imagine the number line as an infinitely long solid tube packed with the ordered rational numbers. A cut will slice the tube into two parts, let's call them A and B, and expose two faces of the tube, the end-points of the sets A and B. By looking at the exposed faces we can read off the number shown. If there is *no* number visible on either face, then we have made the cut at an irrational. In this way Dedekind defined the irrationals in terms of these sets A and B, rather than as sequences. Properties of continuity or limits could thus be formalized in terms of arithmetic rather than vestigial geometric concepts. Bertrand Russell (1872–1970) later noted that the two sets *A* and *B* are each defined by the other, and therefore only one of them is logically necessary, thus an irrational could be defined just in terms of the set *A* (or *B*).

Returning to the question of infinity, Dedekind saw in Bolzano's paradoxes not an anomaly but a definition. He realized that a set is infinite if it is similar to a proper subset of itself; that is, if a subset can be put into a one-to-one correspondence with the set. For example, the set $\{2, 4, 6 \ldots\}$ is a subset of $\{1, 2, 3 \ldots\}$, and can be put in a one-to-one correspondence with it. Two years after Dedekind's book, in 1874, Georg Cantor married, and for their honeymoon he took his bride to Interlaken, where they met Dedekind. In that same year Cantor published one of his most revolutionary papers. He agreed with Dedekind's definition of an infinite set but he also saw that not all infinities were equal.

Cantor started with the fact that any set which can be put in a one-to-one correspondence with a set of natural numbers is countable. This is obvious for finite sets, but Cantor extended the concept of countability to infinite sets. The set of all the naturals is itself 'countably infinite', and any infinite set which can be put in a one-to-one correspondence with it is thus also countably infinite. For example, though the integers seem to disappear into infinity in both the positive and negative directions, they too are countably infinite, as can be seen if we re-order them thus: $\{0, +1, -1, +2, -2 \ldots\}$. Also, just as finite sets are assigned a value known as cardinality (essentially, the size of the set), so Cantor assigned to infinite sets a power. Two infinite sets have the same power if they can be put in a one-to-one correspondence with each other. Now, we saw above that the rationals form a dense set, something which the integers do not, as there is not always a third integer between any two others – for example, there is no integer between 1 and 2. It therefore seemed likely that the set of rationals would have a higher power than the set of integers. However, in 1873 Cantor found this not to be the case. By a skilful arrangement of the rationals he found a method by which they could be put into one-to-one correspondence with the naturals.

With this result it was tempting to think that all infinite sets of numbers do actually have the same power. Cantor showed that this was false by his famous diagonalization argument. He assumed that the reals between 0 and 1 are countable, and could be written down in order and expressed as infinite decimals: for example, 0.2 would be written as 0.199 999 … He then constructed a number which differed from the first in the first decimal place, differed from the second in the second decimal place, and so on. This new number was different from any given number, whose totality had been assumed to be complete, and therefore the real numbers were not countable. The set of reals had a higher power than the set of rationals. Cantor further showed that even the algebraic numbers, which are a far more general class than the rationals, are also of the same power as the naturals. It therefore became increasingly obvious that the continuum of the real numbers was made 'dense' by the existence of transcendentals. In some way, *most* numbers were transcendental.

No one had ever knowingly seen a transcendental number; their actual existence was proved in 1851 by Joseph Liouville. It was only in 1882 that our old friend π was proved to be transcendental, by Ferdinand Lindemann, thereby answering in the negative the centuries-old question of whether it was possible to square the circle using ruler and compass methods. And yet Cantor had more stunning results to come.

In a letter to Dedekind in 1877, Cantor proceeded to prove what Dedekind himself had merely assumed: that the power of the set of points on any line segment is equal to that of any other line segment. Thus a line of unit length has the same number of points as the whole of the number line. Even more surprising was the discovery that this is independent of dimension: the unit line has the same number of points as the unit square or the unit cube – in fact, the same number of points as the whole of three-dimensional space. In what seems almost an understatement, Cantor commented to Dedekind that, 'I see it, but I don't believe it.' Unfortunately, many shared his disbelief.

In 1895 Cantor set down his polished views on his invention of a whole new type of number, the so-called transfinite cardinals. He denotes the countably infinite with the symbol $\aleph_0$ (pronounced 'aleph-null'), and the first non-countable set as $\aleph_1$. There is then a whole infinite sequence of transfinite numbers, each formed as the set of all sets of the previous set. Cantor also proposed that the set $\aleph_1$ is equivalent to the set of real numbers. This is the so-called Continuum Hypothesis, and it still remains unresolved to this day.

Despite such ground-breaking work, Cantor never achieved his ambition of a professorship at the University of Berlin, largely blaming the open hostility of Leopold Kronecker, his old professor from Berlin. Kronecker vehemently opposed Cantor's new branch of mathematics, with sentiments such as, 'God made the integers, all the rest is the work of Man.' The fruitful friendship with Dedekind ceased in 1882 when Dedekind turned down the opportunity to join Cantor at Halle, though their friendship was resumed in 1897 when the two met at a congress. Dedekind seems to have been content in his homely provincial atmosphere, dedicating much of his time to editing the collected works of Dirichlet and Gauss, his former teachers, and Riemann, his much respected contemporary. Cantor remained at the University of Halle. In 1884 he suffered his first mental breakdown; depressions were a recurring theme of his later years. He retired just before the onset of war and died in 1918 in a mental institution in Halle. Antagonism to his work did nothing to ease his mind. But he did live to see his ideas gain rightful recognition as, 'the most astonishing product of mathematical thought', in the words of David Hilbert, one of the leading mathematicians of the early twentieth century. Hilbert continued, 'No one shall expel us from the paradise which Cantor has created for us.' Cantor's work proved fruitful in many branches of mathematics, including a new view of integration theory in terms of the measure of sets, which is where he had started. It also resolved the integration of the Dirichlet function – the answer is *b*.

chapter twenty

# of dice and genes

$$P(x) = \frac{1}{\sigma\sqrt{2\pi}} e^{-(x-\mu)^2/2\sigma^2}$$

$$\int_{-\infty}^{+\infty} P(x)dx = 1$$

◄ This equation represents the Gaussian distribution, better known as the bell-shaped curve showing variability in population characteristics. It is also known as the normal distribution because it is normalised, that is, the whole population is represented somewhere within the distribution, shown mathematically by the integral being equal to unity.

The study of probability as we would think of it today did not begin until the seventeenth century. But investigations into combinations and permutations of objects or events have a longer history, great interest being shown in India, particularly from Jain mathematicians from about 300 BC. The inspiration for the Jains was religious, but for the majority of later writers the spur to understand these processes was the analysis of games of chance – the prediction of possible outcomes, and the establishment of what rules would constitute a fair game. As probability became intertwined with statistics, new methods were developed for analyzing data in both the physical and social sciences. Although it never quite left the gaming-tables, statistics in the Age of Enlightenment was seen as a mathematical way of conducting public policy and ensuring moral and social equity.

The Jain religion developed in India at about the same time as Buddhism, and its mathematical literature dates back to the third or fourth century BC. The Jains showed a particular interest in handling numbers, and a facility for expressing very large numbers. They discussed different types of infinite number and methods of generating them, as well as methods of combining infinite numbers of objects in different ways. They investigated such matters as the different ways of combining the five senses. Interest in permutations is also found in the Vedic literature, with reference to ways of combining syllables in poetic compositions and prayers. In ninth-century Mysore the Jain mathematician Mahavira (*c.* 850) extended this work and gave the now standard rules for combinations and permutations.

The study of combinations and permutations is now known as combinatorial mathematics. A cosmological and mystical use of combinatorials is found in the late-thirteenth-century Catalan philosopher and mystic Ramón Lull (1232–*c.* 1316), but these writings seem to have been largely ignored by mathematicians. The impetus came from the more mundane preoccupation with gambling. Dante's *Divina commedia* mentions the 'game of hazard' which is played with three dice, one player throwing them, the other having to guess their sum. A poem in the thirteenth century, *De vetula,* by the poet known as pseudo-Ovid enumerates the 56 different ways the dice could fall. Both works inspired various commentaries on the mathematical rules of the game, the 'prehistory' of this subject probably ending with Cardano's *Liber de ludo aleae* ('Book of Dice Games'), published posthumously in 1663 but written a hundred years earlier, which deals with how to establish fair bets in both dice and card games.

Probability theory reached a new level of sophistication with the correspondence of 1654 between Blaise Pascal and Pierre de Fermat. They discussed the so-called gambler's problem of points, which concerns the division of stakes between two players when a dice game has to be left unfinished. This problem was tackled by many Italian mathematicians of the Renaissance, including Pacioli, Cardano and Tartaglia, but none of them obtained a full solution. Fermat preferred a method based on listing all possible outcomes and counting the overall winner in each one. The calculations become quite lengthy as the number of games increases, and Pascal preferred a method of expectations. In his *Traité du triangle arithmétique* he elucidated the relationship between the numbers in Pascal's triangle and the required combinations. Each row of the triangle gives the coefficients of the binomial expansion: the third row, for example, gives the numbers 1, 3, 3, 1, which are the coefficients of the expansion $(a+b)^3 = a^3 + 3a^2b + 3ab^2 + b^3$, the number 3 in the second term indicates that there are three combinations giving a^2b, that is, aab, aba and baa. Using the appropriate row in Pascal's triangle, he could thus rapidly decide the division of stakes problem. If player A needs two games to win while player B requires three games, one of the players must win within at least four games; from the row 1, 4, 6, 4, 1 in Pascal's triangle, the stakes should be divided in the ratio of $(1+4+6):(4+1)$ or $11:5$.

These problems were usually discussed in terms of ratios rather than probabilities. The first theoretical discussion of probabilities lying between 0 and 1 occurs in Jakob Bernoulli's *Ars conjectandi*, published posthumously in 1713. He also pointed out that probabilities could be estimated from observed frequencies, and sought to establish an upper limit for the number of trials needed to be 'morally certain' of the estimate of probabilities. Unfortunately, such a stringent condition led to very high values for the number of trials needed: for example, to be 99.9% certain of correctly establishing the ratio of coloured balls in a box would require 25,500 trials. This procedure was refined by Abraham de Moivre, who correctly gave the normal distribution as the limit of the binomial and gave more reasonable numbers of trials needed to approximate the true probabilities experimentally. De Moivre also published a number of editions of his *Annuities on Lives*, which applied these discoveries to pricing annuities and life insurance policies. The impetus for applying probabilistic methods to demographic data actually came from a very different direction. Yet again, we turn our attention to the heavens.

Astronomers attempting to find accurate orbits of the planets had to rely on a number of observations, each of which was prone to a certain error. Each measurement could thus yield a slightly different equation for a planet's orbit, and it was not clear what method should be used to ensure that the most accurate orbit could be computed from a given set of data. Both Kepler and Galileo had struggled with this. The essential idea was to find a curve which minimized the total errors, and in 1805 this was formalized by Legendre in *Nouvelles méthodes pour la détermination des orbites des comètes*, as the method of least squares. It gave a clear justification and a general method which was usable. In 1809 Gauss published his own method in *Theoria motus corporum coelestium*, claiming to have used it since 1795 and thereby provoking a priority dispute with Legendre. It does appear that in 1801 Gauss had used just such a method to calculate the path of the newly discovered asteroid Ceres, using just a few patchy observations made earlier that year. He also showed that the distribution of errors was what is now called the Gaussian or normal curve, and generalized de Moivre's earlier result. Gauss's justification was that this distribution made the average observation the most likely one. Laplace then came up with an even stronger relationship: that whatever the distribution of errors of individual measurements, their averages would tend towards a normal distribution. He also showed that Legendre's least-squares estimates would tend towards the same distribution. Astronomers quickly recognized the usefulness of the method, especially as it was known that errors in astronomical observations were inherent, not merely limited by the instruments but also caused by distortion of starlight as it travelled through turbulent pockets of the atmosphere. In 1812 Laplace published his great treatise *Théorie analytique des probabilités*, in which he synthesized all the developments up to that point; it remained the main text on the subject for a generation.

In a social context, the theory of probability was being seen as 'the calculus of rational behaviour'. In 1814 Laplace said that probability was merely good sense reduced to calculation. Enlightenment mathematicians believed that enlightened individuals acted rationally, and that probability gave the general masses a quantifiable measure by which they could at least imitate the good sense of their betters. The aim was a universal standard of human behaviour; the work on gambling was merely a way of finding the tools for making rational decisions in a world of uncertainties. For example, Laplace and others considered the probability of a jury of a certain size coming to an unjust verdict. But other thinkers did not wholly agree with the rationalist spirit of the French Revolution. John Stuart Mill suggested that reason was better established by observation and experiment than via the pure rational assumptions of probability.

Adolphe Quetelet, a Belgian mathematician and astronomer, supplied a link between the statistics developed for astronomy and social statistics. The idea of the normal distribution was at the root of his idea of the 'average man', around which the characteristics of real people would be distributed, just as the imperfect individual observations of a star would cluster around its true position. Thus deviation from this theoretical norm was seen as a kind of error. He saw it as the business of the state to collect and analyze demographic data so that the 'social physicist' could uncover social laws analogous to physical laws, such as the conservation of a 'moral force'. He sought justification in the fact that the rates of birth, death, crime and marriage appeared to remain stable from year to year, though they might differ from one country to another, thus justifying that each social body had stable but slightly different 'social physics'.

Such social data had been collected since the seventeenth century. In 1662 John Graunt published his *Natural and Political Observations*, based on a statistical analysis of the London Bills of Mortality which were published weekly to act as a barometer for any possible epidemic, and thus give people ample warning to flee the city. In 1693 the astronomer Edmond Halley published a life table based on the Bills of Mortality for the city of Breslau, whose data was more accurate than that available to Graunt. Halley was also able to show that the Government of the day was selling its life annuities too cheaply. But mathematical statistics can properly be seen as a late-nineteenth-century development which brought together the statistical methods of astronomers and the data collection of actuaries.

The biometric tradition was founded by Francis Galton (1822–1911), a cousin of Charles Darwin, who employed statistical methods in the analysis of social data and hereditary characteristics. The main ambition of the so-called eugenics movement was to improve the human species through selective breeding, and statistics was seen as providing a quantitative view of evolution and a guiding hand in its improvement. Galton applied the normal distribution not so much as an 'error curve', but as a measure of variation, realizing from Darwin's theory of evolution by natural selection that biological variability needed analysing in its own right rather than as an evolutionary error from some idealized norm.

I know of scarcely anything so apt to impress the imagination as the wonderful form of cosmic order expressed by the 'Law of Frequency of Error.' The law would have been personified by the Greeks and deified, if they had known of it. It reigns with serenity and in complete self-effacement, amidst the wildest confusion. The huger the mob, and the greater the apparent anarchy, the more perfect is its sway. It is the supreme law of Unreason. Whenever a large sample of chaotic elements are taken in hand and marshaled in the order of their magnitude, an unsuspected and most beautiful form of regularity proves to have been latent all along.

Sir Francis Galton, *Natural Inheritance*, 1889

It was Galton who introduced the notions of regression and correlation. The statistical concept of regression sprang from a study of sweet peas. Galton divided a batch of seeds into seven groups according to seed size. The seeds of the resulting offspring showed the same variability, or variance, in size across the groups. The mean seed size of the whole batch remained constant, but the means of individual groups had shifted away from their parent group towards this population mean. The means thus 'regressed' towards the population mean. By 1885 Galton had understood the phenomenon of regression, and in 1889 he introduced the related idea of correlation. By suitably scaling the two associated variables and plotting their values, Galton discovered a single dimensionless quantity which could represent an index of association between the two variables. This correlation coefficient could vary between +1, perfect positive correlation, to −1, perfect negative correlation. A coefficient close to zero implied no correlation between the variables. The correlation coefficient on its own could not prove any causal connection between the variables, but could justify further experiments to discover such a mechanism.

Galton was to the inheritance of continuous variation what Mendel was to discrete variation, though each worked in ignorance of the other. Gregor Mendel was trained as a mathematician and physicist, and his paper of 1865 suggesting the existence of genes was rediscovered by the biometricians in 1900. It aroused a great deal of controversy, the staunchly Darwinian adherents of the biometric movement largely rejected the notion of genetic material, Pearson thought the idea too metaphysical and could not see how a discrete entity could manifest continuous characteristics. The matter was not resolved until in 1918 Fisher showed that assuming a large enough number of genes the Mendelian model would produce the correlations studied by the biometricians. This was analogous to the discrete binomial distribution tending to the normal distribution with an increasing number of trials.

The philosophical arguments are beyond our scope, but it is important to stress that statistics did not develop as an independent branch of mathematics, but that its developments and the analytical tools were deeply wedded to social concerns. In his later years Galton funded a Professorship in Eugenics (now Human Genetics) at University College, London. The first holder was Karl Pearson (1857–1936), followed by Ronald Aylmer Fisher (1890–1962). In 1901 Pearson and Galton established the journal *Biometrika*, which became the leading statistical journal. It is within its pages that we find not only Galton's theories of regression and correlation, but also Pearson's chi-squared test, developed in 1900, which resolved the problem of assessing whether a theoretical distribution adequately 'fitted' the data to which it was applied. In 1908 W. S. Gossett, a biological scientist working at the Guinness Brewery in Dublin, introduced the t-distribution for small samples. He wrote under the pseudonym 'student' and the t-test is sometimes still referred to as the student-test. Much of Pearson's work was later overshadowed by Fisher, who introduced the analysis of variance, a technique used to test the significance of data from experiments, initially concerned with random block experiments such as are used in agriculture to test fertilizers. The method hinged on mathematically separating any real 'effect' from any random 'error'; if an experiment reveals a real effect, then the method will show the strength of this effect relative to the error.

By the 1920s statistics was seen by mathematicians as a subject of legitimate investigation leading to greater rigour and more subtle methods. Fisher's ideas of experimental design and the analysis of variance were prominent in his book *The Design of Experiment* (1936) and were influential in England and the USA. They radically changed the experimental practice in sciences which deal with variable material not amenable to reproducible laboratory conditions.

chapter twenty-one
war games

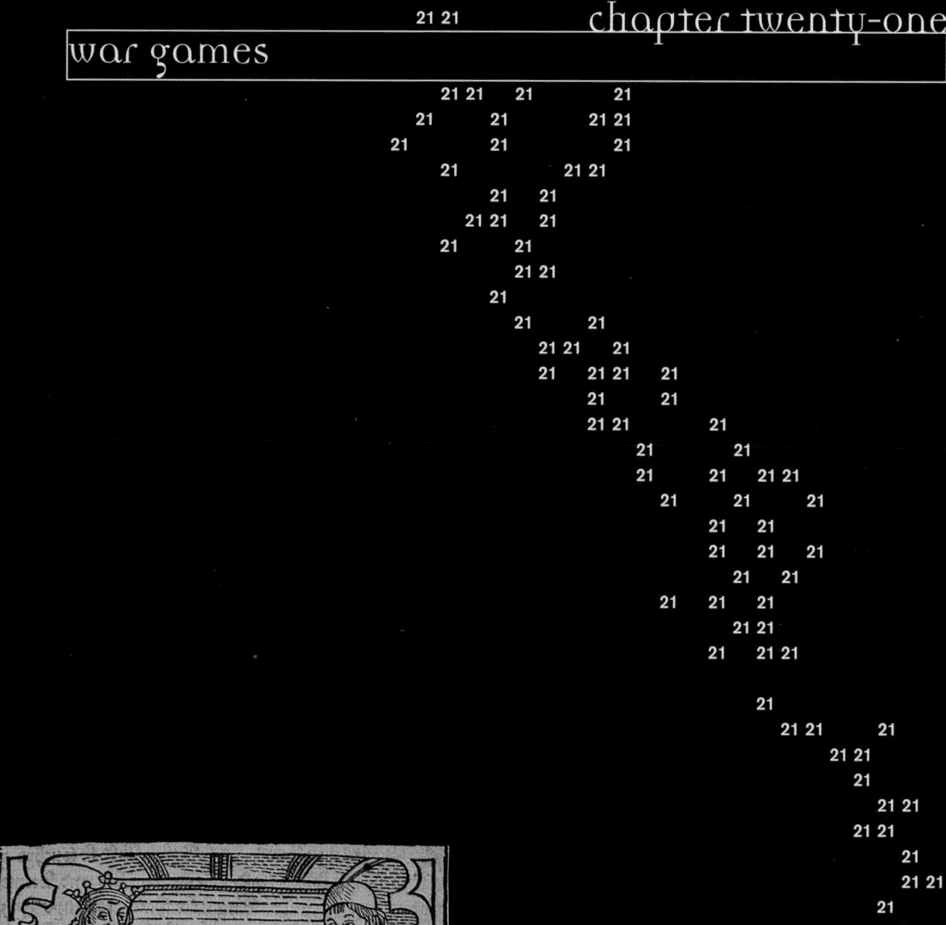

◄ Chess is probably the most popular game of strategy in the world. John Forbes Nash proved that despite its complexity chess has an optimal strategy, the discovery of such a strategy would make chess as trivial as tic-tac-toe.

People have always enjoyed playing games, and every age has had its craze. Most games are a mixture of skill and chance, the truly good player emerging after the vicissitudes of fate have been balanced over many consecutive games. There are some games, however, that leave little to chance – no dice throws, no luck of the draw. These are games based on pure strategy, and their study is the subject of game theory. There are also games that are literally a matter of life or death. As tactical blunders are less costly on a simulated battlefield, military strategists have always turned to war games to refine their skills. It is perhaps no accident that both chess and the Japanese game of *Go* are sublimated war games. It is also no surprise that the first practical application of game theory was in analyzing a new kind of war – potentially the last war.

In the nineteenth century the Prussians invented a game called *Kriegspiel*, literally 'war game'. Played on a board, it was a purely tactical affair, and it grew to be ever more realistic, with an umpire arbitrating over conflict situations with the aid of tables of data from real battles. The Prussian army's military success was largely attributed to their tactical sophistication, developed during *Kriegspiel* simulations. The game was taken up by nations as far away as America and Japan. Germany's defeat in World War I put an abrupt end to the game's mythical status. It was becoming apparent that the rapid development of new weapons and delivery systems meant revising the whole basis of military strategy. The military thus needed mathematicians and scientists not only for developing military hardware, but also for strategic advice – hitherto the domain of generals steeped in military history. This was particularly the case after World War II, and the awareness that two superpowers possessed weapons of mass destruction changed the rules of engagement completely. Board games with cavalry and artillery figures seemed almost prehistoric.

But strategic games continued to be analyzed mathematically in order to gain insights into theories with practical applicability. Émile Borel, French mathematician and minister of the French Navy in the 1920s, wrote his *La Théorie du jeu* in which he analyzed such things as bluffing in poker and the application of the mathematics of games in economics and politics. Borel's influence can be seen in the landmark book *Theory of Games and Economic Behavior*, published in 1944 and written by a Hungarian mathematician, John Von Neumann, and an Austrian economist, Oskar Morgenstern, both then at Princeton. They presented game theory as a possible model for economic interactions. Economists were slow to pick up on the new theory, which in its early years was associated more with military strategies.

János von Neumann (1903–57), later known as John Von Neumann, was born in Budapest and showed phenomenal calculating abilities from an early age. In 1921 he gained one of the limited places reserved for Jews at the University of Budapest, and was awarded a doctorate in 1926 for a thesis on game theory, despite never having attended a lecture. Instead he spent the intervening years in Berlin and Zurich studying chemistry, his father's preferred choice of career subject, while continuing his mathematical studies with mathematicians such as Hermann Weyl and George Pólya, later studying with David Hilbert at Göttingen. In 1930 he went to Princeton, and in 1933 became one of the five original mathematicians to join the newly founded Institute for Advanced Studies at Princeton, where he was to spend the rest of his working life. He resigned from his posts in Germany when the Nazis came to power and decided to settle in America, not as a refugee but because he thought there were more opportunities there. From 1940 he held numerous advisory positions giving scientific advice on military matters, he worked at Los Alamos on the quantum mechanics for the production of the atomic bomb, and in 1955 he was appointed to the Atomic Energy Commission. From those days in Zurich, Pólya recounts that, 'Johnny was the only student I was ever afraid of. If in the course of a lecture I stated an unsolved problem, the chances were he'd come to me as soon as the lecture was over, with the complete solution in a few scribbles on a slip of paper.' He died in 1957 from cancer, and friends tell of his despair at losing his mental faculties after a lifetime spent cultivating them. His most remembered work was on game theory, quantum mechanics and computing.

The simplest type of game is the two-player two-strategy zero-sum game – a game in which two perfectly rational players are each intent on winning, and in which the total utility is zero, i.e. one player's gain is the other's loss. An amusing example is known as the 'cake division' game. A scenario replayed in many a household is the division of a cake between two children so that neither feels that the other has the larger piece. The solution is a two-step process: one child cuts the cake in half, and the second child has the first choice. Both children would like the largest piece, but on the rational assumption that each child recognizes the greed of the other there is an optimal solution. The first child must cut the cake in the fairest possible way because if one piece is much larger then the second child will no doubt choose that one in preference. The so-called minimax theory expounded by Von Neumann says that there will be a 'saddle point' or optimal solution in this case where both players will be content. The theory was expanded to include more than two players, and as the number of players increases the implementation becomes ever more intractable. Much of the book discusses games in terms of payoff tables for the players and as the number of players increases the tables become ever larger, requiring large matrices for their calculation.

In the 1940s John Forbes Nash extended Von Neumann's theory to non-zero-sum games. An example of such a game is the stock market: there may be winners and losers among the players, but the total pot of money also changes as the market capitalization increases. Nash discovered that non-zero-sum games also had an equilibrium solution. He was born in 1928 in West Virginia, graduated from the Carnegie Institute of Technology and gained a doctorate from Princeton in 1950 with a thesis on non-cooperative games. While studying for his doctorate he wrote a paper which much jointly won the 1994 Nobel Prize in Economic Science. From 1951 he taught at the Massachusetts Institute of Technology, where he also did ground-breaking work on geometry, Riemann manifolds and Euclidean space. In 1959 this most promising of young mathematicians was struck down by schizophrenia. His experiences and recovery in the mid-1970s were personally described at the World Congress of Psychiatry in 1996. He continued to produce outstanding work, even during his stays in hospital, in areas such as geometry, topology and differential equations. He continues to work on the geometry of space.

Nash's work showed that there are scenarios in which the optimal outcome does not result from what appears to be the obvious courses of action. A famous illustration example is the so-called prisoner's dilemma, devised by Melvin Dresher and given this interpretation by Albert Tucker in a lecture to psychology students. The scenario has changed with the retelling, but in its original form Tucker explains that two men have been arrested for a violation of the law and placed in separate cells. If one confesses he will receive a reward and the other will be fined; if both confess, then both will be fined; and if neither confesses, they will both be released. The essence of the dilemma is that the optimal outcome is for both to keep quiet and thus both be released, but the fear that such a strategy may backfire if the other man confesses may well lead both to confess, in which case both men will be fined. It is with such strategic games and scenarios in mind that applications were sought in the realms of negotiations, be they military, business or personal. Experimentally it was found that people had a keen sense of the theoretically optimal solution, and an occasional defection would result in immediate retaliation from the other party, a tactic known as tit-for-tat.

There are games in which there is an optimal strategy, and once this has been found then the game becomes essentially trivial. For example, noughts and crosses (tic-tac-toe) is a popular children's game, but once the strategy has been understood and each player plays rationally, then every game is drawn and interest wanes.

➤ This is the starting position of Karl Sims's *Evolved Virtual Creatures* (1994). The creatures affectionately known as 'blockies' compete in a game to gain control of a green cube. They are the results of a simulated Darwinian process where their bodies and behaviours are evolved to perform certain tasks.

Nash proved that even chess has an optimal strategy, but the game is so complex that the optimal strategy has not yet been found, even to the point where it is not clear whether the outcome will be a draw or a win for white. If an optimal strategy is ever found then chess will become as trivial as noughts and crosses. Was there an optimal strategy for nuclear weapons? For a few brief years America was the sole nuclear power, but the fear that Russia would build a nuclear arsenal drove some thinkers such as Von Neumann, and even Bertrand Russell, to advocate an immediate first nuclear strike against Russia and the setting up of a world parliament to enforce global peace. This was not implemented, and the policy soon changed to one of deterrence and mutually assured destruction (MAD). Such strategies were largely developed at the secretive think-tank the RAND Corporation.

The RAND Corporation was founded in 1945 with leftover defence funding at the end of the war, initially as part of the Douglas Aircraft project, and in 1948 it was formally constituted as a non-profit organization with funding from military and business sectors. It is the archetypal think-tank, whose brains were there to 'think the unthinkable'. RAND stood for 'research and development', most of which was focused on national strategies in a nuclear world. All the mathematicians in the USA during the 1940s and 1950s mentioned so far worked there at one time or another. Nash introduced them to a host of games, including *Kriegspiel*. The logistics of war were studied with precision, and failsafe mechanisms were put in place to avert any accidental strikes. With the fear on both sides of the growing stockpile of weapons, a tit-for-tat strategy seemed unlikely – the nuclear game was one that could be played once and once only. The strain of superpower politics for two generations came to tell on the world's population and their leaders. Thinking the unthinkable meant that neither side did the undoable.

RAND operated more like a university than a military agency, giving its brains the freedom to pursue their invariably peculiar lifestyles, and the building was open around

➤ At each generation, the most successful creatures survive and reproduce, with mutations and recombinations applied to their offspring's virtual genetic material. When the cycle of selection and variation continues for many generations, successful strategies are automatically 'invented' with no human designer involved. The green cube is now within reach.

the clock. RAND had a flourishing publishing arm. One of its most popular books, published in 1954, was *The Compleat Strategyst* by John D. Williams, a popular account of the applications of game theory for the layman, incorporating the kind of gallows humour common at the corporation. Many other think-tanks now exist, spawned by RAND's success, but probably none have had this intense mixture of mathematicians as permanent staff locked in abstract thought.

The terminology used in such strategic games is of cooperation and defection. Game theory was later much criticized for its cynicism in treating people as wholly self-serving, but subsequent studies have shown that people's real-life strategies do reflect their perceptions of relative gain. In a zero-sum game a draw would leave each player no better off than at the start, but in a non-zero-sum game such as the stock market, winning and losing are relative, and the game is more about maximizing one's own winnings than beating an opponent. Cooperation thus becomes more common if both parties gain from the transaction. Although initially slow to develop, game theory is now an integral part of the analysis of the market economy. A recent use has been in the global phenomenon of the auctioning of public utility licenses to private businesses, thereby gaining much needed revenue and opening up new markets for development. The whole global marketplace is a shifting scene between collaborations and competition – a world of game theory.

I propose to consider the question, 'Can machines think?' ...

The new form of the problem can be described in terms of a game which we call the 'imitation game'. It is played with three people, a man (A), a woman (B), and an interrogator (C) who may be of either sex. The interrogator stays in a room apart from the other two. The object of the game for the interrogator is to determine which of the other two is the man and which is the woman ... It is A's object in the game to try and cause C to make the wrong identification ... In order that tones of voice may not help the interrogator the answers should be written, or better still, typewritten. The ideal arrangement is to have a teleprinter communicating between the two rooms ... The object of the game for the third player (B) is to help the interrogator ... she can add such things as 'I am the woman, don't listen to him!' to her answers, but it will avail nothing as the man can make similar remarks.

We now ask the question, 'What will happen when a machine takes the part of A in this game?'. Will the interrogator decide wrongly as often when the game is played like this as he does when the game is played between a man and a woman? These questions replace our original, 'Can machines think?'

Alan Turing, *Can a Machine Think?*, 1950

chapter twenty-two

maths and modern art

◀ Giacomo Balla, *Abstract Speed – The Car has Passed*, 1913. Balla was a signatory to the Futurist Manifesto of 1910 which declared 'that universal dynamism must be rendered in painting as dynamic sensation.'

The twentieth century has seen an explosion of scientific discoveries and technological advances across the whole range of physical, biological and human sciences. During the Enlightenment it was believed that our accumulated knowledge would give us unlimited power over nature and release us from our earthly toils. The artistic responses to these developments had not always been positive, witness William Blake's repudiation of a Newtonian clockwork universe. In the early twentieth century our view of the universe changed radically – with relativity and quantum mechanics the universe regained its mystery and magic. However, as scientific and political developments collided in two world wars, there were indeed many good reasons to re-evaluate our place in the universe. It is hoped that in the future we can develop our wisdom in proportion to our knowledge.

I have looked at the role of mathematics in these developments in other chapters. Here I shall concentrate on the influence of mathematics, sometimes hand in hand with the new physics, on general culture and the arts. The arts are often the most public expression of philosophical shifts and of the personal reactions of artists to the changing technological environment. It is certainly not the case that mathematics was the only influence or even the main influence in various cultural movements, but it is interesting to look at those areas in which mathematics has played a unique and important role. The very use of mathematical terms within artistic discourses has shown that artists have taken on board the language and ideas of mathematics and transformed them through the prism of artistic life.

Many of the new art movements that sprang up in the first two decades of the twentieth century embraced the language and ideas of the new geometries developed by mathematicians. Painting and sculpture are, by their very nature, concerned with artistic expressions in two and three dimensions respectively. But these are limitations to a full representation of the world and human existence. What did the new geometries contribute to the new modes of expression?

During the Italian Renaissance, the mathematics of perspective allowed a more realistic representation of three dimensions on a two-dimensional surface. Perspective broadened the language of painting, and artists were quick to learn the new rules. Later, they consciously broke the same rules for visual and aesthetic effect. In the twentieth century, Cubism, Futurism and Surrealism all came to terms with concepts from the new geometries, such as non-Euclidean geometry and multi-dimensional space, specifically the fourth dimension. In the early years of the century the new geometries were an important influence on individual artists more than on each movement as a whole. By the end of the 1920s the temporal fourth dimension of Einstein's theory of relativity had gained prominence, but before then there had been much investigation of a spatial fourth dimension. The mathematical revolutions in geometry took place in the middle of the nineteenth century, with the discovery of non-Euclidean geometry independently by Lobachevsky and Bolyai around 1830 (Chapter 16). In 1854 Bernhard Riemann published

▲ David Bomberg, *In the Hold*, 1913–14. A grid structure superimposed on a scene of human figures rendered as rectilinear shapes creates a picture simultaneously fractured and yet highly dynamic.

his seminal 'On the Hypotheses which Lie at the Foundation of Geometry', which set the stage for mathematical research on multi-dimensional spaces and physics experiments on the true geometry of space.

Euclidean geometry was now only one of many possible geometries. The actual geometry of space itself was and continues to be a matter for mathematicians and physicists, but at the same time the geometry of perceptions and representations began to be explored by artists. If we first look at the extension of the idea of three spatial dimensions to a fourth one, we immediately come across the problem of representation. The standard analogy is the one employed so successfully in Edwin Abbott's book *Flatland* (1884), which is to look at the perceptions of two-dimensional beings living in a 'flat land' when they are visited by a three-dimensional object. Such a view was illustrated by Claude Bragdon in a number of books including *Man the Square: A Higher Space Parable* (1912). The idea is to try to gain an intuitive grasp of the whole object by taking successive slices or cross-sections through it. Thus, as far as painting as a medium was concerned, to grasp the whole object, whether it exists in three or in four dimensions, required a sequence of slices through it or multiple images of it from multiple perspectives. This is indeed one of the ways in which the Cubists represented their subjects.

Perspective was seen as limiting, and rejected as providing too narrow a view on objects. The philosopher Immanuel Kant's distinction between the perception of objects and the objects themselves also inspired Cubism's multifaceted forms. In fact the fourth dimension was being given a number of formulations beyond the strictly mathematical and spatial: to some it was the Platonic realm of ideals, of mysticism or of the irrational. In short, the fourth dimension freed the artist to explore reality beyond a three-dimensional perspective. This freedom was taken up not only by the Cubists, but also by the Italian Futurists, whose intellectual manifesto of 1909 was part political and part artistic, promoting modernity, industrialism and technology. Artists such as Umberto Boccioni, Gino Severini and Giacomo Balla expressed the dynamism of the fourth dimension.

Probably the most influential mathematician in the public realm in France at the time was Henri Poincaré, a respected intellectual whose writings went beyond mathematics to politics, education and ethics. In 1906 he was made President of the Académie des Sciences, and his popular writings certainly put physics and mathematics in the public arena. His philosophy on the relativity of knowledge and his focus on the creative side of mathematical activity, including the role of the subconscious incubation of difficult problems in contrast to a purely logical line of thought, were powerful influences in the early twentieth century. Perhaps just as influential within Cubist circles was a little-known mathematician, Maurice Princet, an actuary and amateur painter who explored the mathematics of non-Euclidean geometry with the artists Jean Metzinger and Juan Gris.

Let us imagine for a minute any three-dimensional body, an African lion for example, between any two moments of his existence. Between the lion LO, or lion at the moment T=O, and the lion LI or final lion, is located an infinity of African lions, of diverse aspects and forms. Now if we consider the ensemble formed by all the points of lion to all its instants and in all its positions, and then if we trace the enveloping surface, we will obtain an enveloping super-lion endowed with extremely delicate and nuanced morphological characteristics. It is to such surfaces that we give the name lithochronic.

Oscar Dominguez, 'La Petrification du temps,' 1942

In 1905 Albert Einstein, still a patent officer, first wrote on the special theory of relativity; in 1916, now a professor, he published on the general theory. By the end of the 1920s the fourth dimension as a spatial dimension was almost wholly replaced by the idea of a fourth dimension that was temporal. Time and hence motion became of prime concern to artists such as Marcel Duchamp and Umberto Boccioni, as in his sculpture *Unique Forms of Continuity in Space* (1913), as well as to Frank Kupka and the abstract art of Kasimir Malevich.

Cubism was founded by Pablo Picasso and Georges Braque, Picasso's *Les Demoiselles d'Avignon* of 1907 being the first Cubist painting. The most fertile period of Cubism came to an end in 1922, its practitioners by then having diverged from the earlier, unified style. Although it had been regarded as a coherent school of art, there had always been differences in the underlying philosophy and practice. Picasso himself seems to have been little influenced by mathematical ideas, stating that his own

➤ Marcel Duchamp, *Nude descending a staircase, No.2*, 1912. Influenced by the 'geometric chronophotography' of Marey and the 'cinematography' of Muybridge from the 1880s, Duchamp creates a more plastic cubism by representing a temporal fourth dimension on a two-dimensional canvas.

influences were the shifting perspectives of Cézanne and the structures of African art and sculpture. Braque himself was much concerned with geometric representations, and in fact it was he who coined the term 'Cubism'. There was indeed also a strand of continued interest in more traditional geometric considerations of perspective and structure. In 1912 there took place in Paris the highly influential exhibition entitled *Section d'Or*, or 'Golden Section', a reference to the classical proportion often found in architecture and art. At this time, artists such as Gris and Jacques Villon approached a purely abstract and geometric form of Cubism shorn of any representational aspects.

The influence of non-Euclidean geometry on early twentieth-century art is more difficult to quantify than that of a fourth dimension. The problem may stem from the difficulty in representing non-Euclidean spaces. The Italian mathematician Eugenio Beltrami (1835–1900) made a physical model of the pseudosphere as a representation of Lobachevskian geometry. The mere knowledge of its existence may have been enough for non-Euclidean geometry to fire the artistic imagination; perhaps its formal mathematical character made it a less fertile concept than the artistic freedoms offered by a fourth dimension. Painters such as Duchamp were influential, but in the minority, in persuading artists to study mathematics and science. Considerations of non-Euclidean geometry, however, influenced the founder of Dada, Tristan Tzara, and the Surrealists.

In 1936 the painter Charles Sirato published the *Manifeste dimensioniste*. Citing the theories of Einstein as one of its inspirations, the manifesto declares that, 'animated by a new conception of the world' the arts have evolved into a new dimension. Painting was to leave the plane and occupy space, thereby leading to spatial constructions and multi-media installations. It urged 'sculpture to abandon closed, immobile, and dead space, that is to say, the three-dimensional space of Euclid, in order to conquer for artistic expression the four-dimensional space of [Hermann] Minkovski.' The manifesto was signed by an impressive selection of leading artists. The declaration allowed for both main interpre-tations of the fourth dimension, that is, as a spatial and spiritual dimension and as time.

By and large, however, few painters after the 1930s were demonstrating an open interest in either the fourth spatial dimension or non-Euclidean except for the Surrealists. André Breton found the new geometries ideally suited to his arguments for a new 'surreality'. Although Breton's Surrealist theory was based largely on Freud's analysis of the unconscious, higher dimensions infused it too, four-dimensional space-time being combined with higher dimensions of the irrational or subconscious. We can see this interest in the titles of some of the works, such as Max Ernst's *Young Man Intrigued by the Flight of a Non-Euclidean Fly* (1942), and also in their content, as for example in Salvador Dali's famous 'soft' watches in, among other works, *The Persistence of Memory* (1931), and the hypercube, the four-dimensional counterpart of the cube, in his *Crucifixion (Corpus Hypercubicus)* of 1954. The most scientific of the Surrealists

▲ Salvador Dali, *The Sacrament of the Last Supper*, 1955. Euclidean geometry continued to inspire artists and here the last supper takes place within a regular dodecahedron, the platonic symbol of the whole universe.

was Oscar Dominguez who, working in sculpture, was fascinated by the life of objects in time. His ideas of 'lithochronic surfaces' seem very close to Boccioni's sculptural works. By 1939 Dominguez had produced a series of highly spatial 'cosmic' paintings whose polyhedral forms have been compared with geometric models exhibited in the Henri Poincaré Institut and photographed by Man Ray for the 1936 Surrealist Exhibition. A truly mathematical, and aesthetic, representation of non-Euclidean geometries had to wait for the power of computers to become a reality.

The new multi-dimensional and non-Euclidean geometries, which started life as abstract mathematical theories, were not only being employed in the new physics but were an inspiration to artistic and philosophical movements which sought to overthrow established modes of thought. In the world of art these expressions took on a variety of forms, ranging from the spiritual to the anarchic – and even both simultaneously. The abandonment of Euclidean geometry as *the* spatial paradigm meant that space was created for new perspectives on life, the universe and everything.

The new artists have been violently attacked for their preoccupation with geometry. Yet geometrical figures are the essence of drawing. Geometry, the science of space, its dimensions and relations, has always determined the norms and rules of painting.

Until now, the three dimensions of Euclid's geometry were sufficient to the restiveness felt by great artists yearning for the infinite.

The new painters do not propose, any more than did their predecessors, to be geometers. But it may be said that geometry is to the plastic arts what grammar is to the art of the writer. Today, scientists no longer limit themselves to the three dimensions of Euclid. The painters have been quite naturally, one might say by intuition, to preoccupy themselves with the new possibilities of spatial measurement which, in the language of the modern studios, are designated by the term: the fourth dimension.

Regarded from the plastic point of view, the fourth dimension appears to spring from the three known dimensions: it represents the immensity of space eternalizing itself in all directions at any given moment. It is space itself, the dimension of the infinite; the fourth dimension endows objects with plasticity. It gives the object its right proportions on the whole, whereas in Greek art, for instance, a somewhat mechanical rhythm constantly destroys the proportions.

Greek art had a purely human conception of beauty. It took man as the measure of perfection. But the art of the new painters takes the infinite universe as its ideal, and it is to this ideal that we owe a new norm of the perfect, one which permits the painter to proportion objects in accordance with the degree of plasticity he desires them to have ...

Finally, I must point out that the fourth dimension ... has come to stand for the aspirations and premonitions of the many young artists who contemplate Egyptian, Negro and oceanic sculptures, meditate on various scientific works, and live in the anticipation of a sublime art.

Guillaume Apollinaire, from 'La peinture moderne', *Les Soirées de Paris*, April 1972

chapter twenty-three

machine codes

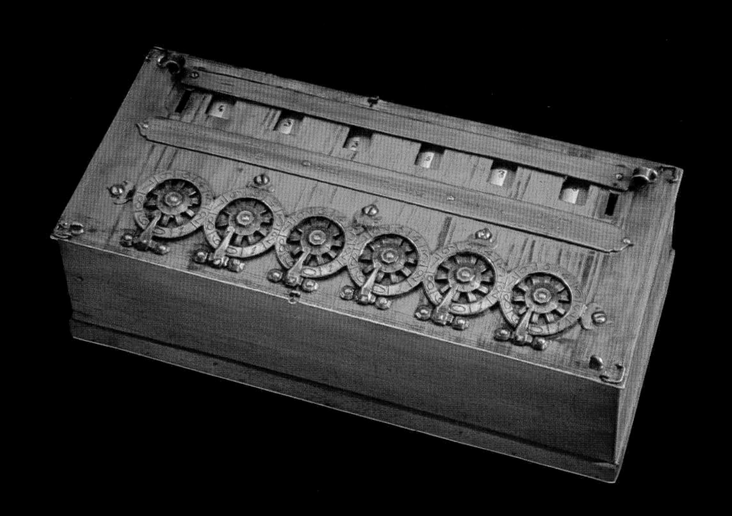

◄ One of the first calculators was invented by Blaise Pascal in 1642. Addition was performed by turning the wheels with a stylus, but other operations proved rather cumbersome.

▲ Medieval tally sticks such as these were used by the English Exchequer as late as 1826, when they upgraded technology to pen and paper. The amounts paid to the Exchequer were registered with notches cut into the side of the tally stick, which could then be separated into two pieces, one for each party.

In the history of mathematics there have been a number of double strands whose relative importance has ebbed and flowed, like the relationship between arithmetic and geometry, or between pure and applied mathematics. Another pair of dynamic opposites can be seen in algorithmic and 'analytical' mathematics. The latter is more concerned with the underlying structure and 'beautiful theorems', whereas the former aims to understand the procedures required to reach practical solutions. We have seen, for example, that various methods, or algorithms, have been used to find irrationals such as $\sqrt{2}$ in a variety of number systems. Investigating which procedure is most efficient, in terms of reaching the required level of accuracy in the smallest number of steps, is a major concern of algorithmic mathematics.

Originally, the term 'algorithm' referred to performing calculations with Hindu-Arabic numerals, as distinct from calculations on the abacus or counting-board. As the use of the abacus declined in Europe, and as calculations became ever more labour-intensive, the desire for mechanical calculators grew. Mathematicians such as Pascal, Descartes and Leibniz in the seventeenth century dreamt of a universal language in which all mathematical problems could be encoded, and which would generate methods of solution which could be implemented mechanically. They themselves built a variety of mechanical calculators. Leibniz's vision of a universal calculus went beyond the differential and integral calculus to include formal rules which could decide questions of science, ethics and law. The power of any efficient algorithm can be greatly magnified by employing calculating machines, but it was not until the twentieth century that the software and the hardware came together.

Charles Babbage (1791–1871) produced the first design for his 'Difference Engine' in 1819, and had a working prototype built by 1822. He hoped it would improve both the speed of production and the accuracy of much needed tables of figures, such as for logarithms. The British government gave its backing to the construction of a full-scale machine capable of producing tables of unprecedented accuracy for actuarial, administrative and scientific use. But by 1834 the project was well over-budget and behind schedule. Although Babbage's intellectual and financial commitment was never in doubt, the government procrastinated over further funding. By then he had shifted his attentions to the design of the 'Analytical Engine', the real precursor to the modern computer. A key feature was the separation of the store, which held numbers during computation, and the mill, which performed the arithmetic operations. The input and output were encoded on punched cards, as was the control device which controlled the program. Output directly to a printer was also envisaged, as was a steam engine so that the whole operation could be automated. But the Analytical Engine was never built, and in 1842 the government decided to cease funding the Difference Engine. Babbage's jaundiced views of British science were more than confirmed. As an undergraduate he had co-founded the Analytical Society, whose main aim was to bring the teaching of mathematics at

Cambridge up to the standards of the Continent. In 1830 he wrote a blistering attack on the state of British science, largely blaming the insularity of the Royal Society, and this led to the foundation of the British Association for the Advancement of Science. Unfortunately, the cost-benefit analysis of Babbage's ventures meant they were to languish for nearly a hundred years.

As Babbage had foreseen, the development of computers sprang largely from the need to speed up operations performed by the mechanical calculators then in use. The first automatic calculating machine was built at Harvard with funding from IBM around 1940. The first electronic programmable computer was the Colossus, built in 1943 with the cooperation of Alan Turing and John von Neumann as part of Britain's code-breaking work at Bletchley Park. However, the design that was to influence future computer architecture was the ENIAC (Electronic Numerator Integrator And Computer), built at the University of Pennsylvania around the same time. Von Neumann had hoped that the ENIAC, originally designed to compute ballistics tables, would be able to carry out some of the calculations required by the Manhattan Project, but instead found himself proposing a new design for a stored-program computer called EDVAC (Electronic Discrete Variable Automatic Computer). The new structure was to have five major components: an input, an output, a control unit, a memory and an arithmetic unit. The stored-program computer is so called because the program is held in memory with the numerical data, while the control unit executes a sequence of instructions. The first practical computer of this type was built in 1949 in the UK, known as EDSAC (Electronic Delay Store

➤ The Science Museum, London, constructed in 1991 the first complete Difference Engine in honour of the bicentenary of Charles Babbage's birth. Consisting of some 4000 parts it weighs over 2.5 tons. This was envisaged by Babbage to be a fully automated computer, with printed output and powered by steam engine.

▲ A re-creation of the Colossus code-breaking computer at Bletchley Park (1997). It was the first electronic programmable computer in the world and helped cryptographers break the German Lorenz code during WWII.

Automatic Calculator). Further machines were built in both the USA and the UK, and the stored-program design was common by the 1960s. The introduction of semiconductor components to replace cathode-ray tubes led to increases in speed and reliability. Despite similarities with Babbage's designs, these developments took place in complete ignorance of his earlier work.

As we can gather, the invention of computers was motivated by very practical concerns, whether in business and administration, cryptography, or solving equations in mathematical physics. The stored-program computers had separated the hardware from the software. Initial work on programs, algorithms to execute the appropriate calculations, stemmed not from practical concerns, but from logicians looking at formal systems.

Ordinary arithmetic is a common example of a formal system. It has a well-defined set of symbols, and procedures for manipulating these symbols so as to get from a problem to its solution. The symbols themselves have no meaning except with reference to the rules of the formal system. For example, if I want to verify that ABmBAeBEB, there are various methods or algorithms I can use to perform the calculation. It will probably be easier to see the calculation as $12 \times 21 = 252$, but it does illustrate that the actual symbols used are not important – what matters is that the truth of a proposition, in this case a calculation, can be proved from established axioms. Indeed, in a reversal of the common use of letters to denote functions, numbers need not even signify quantities, but can also act as operators. This is a crucial factor in the shift from computational devices designed for specific types of problem to a general, all-purpose computer. In a modern computer any command, for example to display a red pixel at a specific location on your monitor, is essentially a string of numbers. In fact, a whole program, once encoded into binary, is essentially one (very large) number. The underlying simplicity of computers is often overlooked because of the remarkable progress in the speed and power of these machines.

Kurt Gödel (1906–78), in his 1931 paper 'On formally undecidable propositions of *Principia mathematica* and related systems', gave a particular method of assigning a unique number to every proposition expressible within the formal system. Even the proof of a proposition's truth-value could be expressed as a unique string of natural numbers, to the extent that given a string of such basic symbols it is possible to decide which are meaningful and which are not. The first of two classic results in this paper, known as the Incompleteness Theorems, is that an axiomatic system as basic as the arithmetic of whole numbers contains propositions which cannot be proved to be either true or false. This is in some ways analogous to the linguistic 'this sentence is false' dilemma. The existence of such undecidable propositions showed that the programme of axiomatization of mathematics embarked upon by Bertrand Russell and Alfred North Whitehead was not feasible. Gödel also disappointed David Hilbert, who sought to reach the goal of a complete and self-consistent arithmetic, that is, one with no internal contradictions. Gödel's second result showed that the opposite was the case, that if the system is consistent it cannot actually prove its own consistency from within itself. In short, we now say that arithmetic is incomplete. After such an apparent body-blow it suffices to say that mathematicians abandoned the quest for a grand unified mathematics and focused instead on how different forms of axiomatization lead to different systems. The whole point of having a mathematical language is to enable us to answer questions, so discussions stressed the process by which the truth of mathematical propositions is decided. Mathematicians were now talking about computability rather than decidability.

In parallel with the focus on algorithms, came a generalization of the concept of a function. In the most general sense, a function f is an arbitrary correspondence between mathematical objects. A function is said to be computable if there exists an algorithm which produces as its output f(*x*) for an input *x*, and if this is possible for every value of *x* for which f is defined, the so-called domain of f. It was only when pathological functions were constructed at the end of the nineteenth century did it occur to mathematicians that a function might not be computable. Attention thus shifted to computational algorithms. It had been fairly clear whether a given algorithm accurately computed the required function, but if no algorithm could be found, and one then proceeded to prove that there was no such algorithm, then one needed a precise definition of what an algorithm actually was in order to prove that none existed. Gödel's paper also contained ideas on recursive functions, where the required function is reached by a succession of well-defined intermediate functions. These notions proved to be very fruitful in defining computational algorithms. In 1936 Alonzo Church in Princeton and Alan Turing in Cambridge both published independently their notions of computability, and Turing then proceeded to show that the two notions were essentially equivalent. Turing's definition of an algorithm was based on a machine model of computation, swiftly named by Church a 'Turing machine'. The pressing question of whether there was an algorithm that could determine whether a given formula was true or not was answered in the negative. These results, coupled with Gödel's, effectively ended the hope that a computer would one day be able to sort out the truth or falsehood of all mathematical propositions. However, the focus on algorithms gave rise to developments in software, which heralded a new era in mathematical physics. Computational mathematics took up the challenge of old problems in dynamics, such as the stability of the Solar System, as well as turning its attention to biological systems and the complex dynamics of life itself.

The Analogy principle. All computing automata fall into two great classes in a way which is immediately obvious and which ... carries over to living organisms. This classification is into analogy and digital machines.

Let us consider the analogy principle first. A computing machine may be based on the principle that numbers are represented by certain physical quantities ... currents may be multiplied by feeding them into the two magnets of a

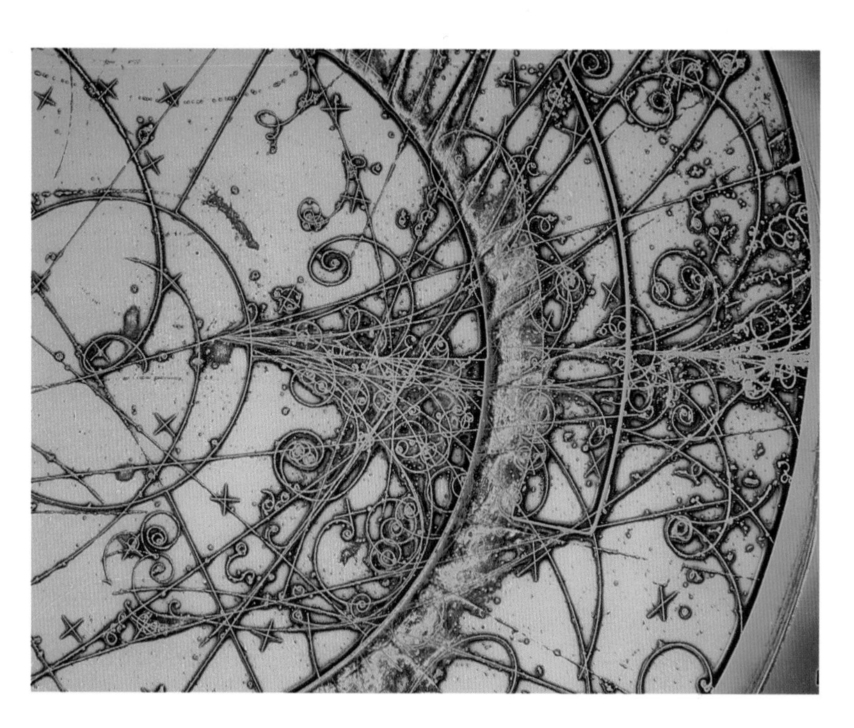

▲ Particle tracks from the Big European Bubble Chamber, CERN. Computers have reached such speed and power that they are helping physicists to explore the fundamental forces of nature. Somewhat less power was needed to produce the book you are now reading!

dynamometer, thus producing a rotation. This rotation may then be transformed into an electrical resistance by the attachment of a rheostat; and, finally, the resistance can be transformed into a current by connecting it to two sources of fixed (and different) electrical potentials. The entire aggregate is thus a 'black box' into which two currents are fed and which produces a current equal to their product ...

The Digital Principle. A digital machine works with the familiar method of representing numbers as aggregates of digits. This is, by the way, the procedure, which all of us use in our individual, non-mechanical computing, where we express numbers in the decimal system. Strictly speaking, digital computing need not be decimal. Any integer larger than one may be used as the basis of a digital notation for numbers. The decimal system (base 10) is the most common one, and all digital machines built to date operate in this system. It seems likely, however, that the binary (base 2) system will, in the end, prove preferable, and a number of digital machines using that system are now under construction.

John von Neumann, *The General and Logical Theory of Automata*, 1948

chapter twenty-four

chaos and complexity

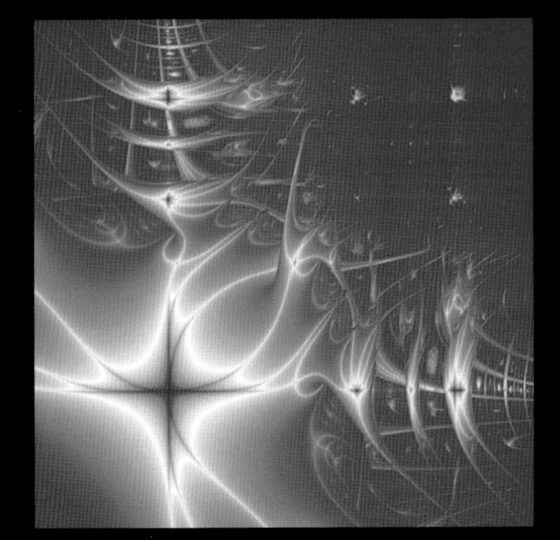

◄ A computer generated image representing the stability of a dynamical system to small perturbations using the so-called Lyapunov exponent. The sharply-defined curves represent regions of super-stability while crossing regions indicate where different attractors compete to dominate the system's behaviour. The background colour represents regions of chaos.

From the beginning of the nineteenth century, mathematics was pursued as an analytical and logical subject; by the end of that century it had produced a menagerie of mathematical monsters, such as continuous functions without tangents. In dynamics, the three-body problem – the test case for the stability of the Solar System – still had no stable solutions, and Henri Poincaré had discovered a very intricate structure while analysing a partial solution of it. A change of viewpoint from the analytic to the geometric showed mathematicians that what appeared to be a horrendous mess had many similarities with the apparent mess that is the actual world. The mathematical monsters turned out to be guarding an Aladdin's cave of new and wonderful mathematical objects. Entrance to this world was through the use of computers, which became the laboratories of the new algorithm-based mathematics. In turn, the discoveries they made possible bolstered analytical understanding and led to the recognition that the 'simple' systems mathematicians were used to are but the tip of a very large iceberg indeed.

The name synonymous with the field of fractal geometry is that of Benoît Mandelbrot, currently a professor at Yale University and an IBM Fellow Emeritus. His interest in what he later named fractals, which began in 1951, culminated in 1977 with his book *The Fractal Geometry of Nature*. Much has been written about fractals, and their rococo-like images are well known. The idea of an 'attractor' was well known in dynamics – for example, the orbit of a planet is an elliptical attractor, and though perturbations take place they are held in check within certain bounds. In the solution of polynomials by numerical methods, if the iteration converges to a solution then that solution is an attractor. Sometimes a root that is known to exist graphically cannot be reached by an iterative process; such a root is called a 'repeller'. But in a chaotic system such as turbulent airflow, the attractor is a fractal, and is known as a strange attractor.

Once we learn how to look at things in the right way, we find chaotic behaviour in the simplest of situations. The logistic difference equation $z = \lambda z(1 - z)$ is a simple quadratic with only one changeable coefficient, denoted by λ. The equation has two roots, as we would expect for a quadratic, but if we use an iterative procedure we discover some amazing properties. For most values of λ the iteration 'blows up', and diverges to infinity. But if we start from $\lambda = 1$ and slowly increase the value, we find that the iteration does not diverge, nor does it converge to a single value: instead, it oscillates between a number of values. At some point the system goes chaotic, with the sequence of numbers seemingly jumping around wildly. If we now expand into complex numbers, the real-line segment fans out to reveal a fractal structure. With a simple transformation the difference equation becomes another quadratic, $z = z^2 - m$. The iterative process is simple but very tedious by hand. With the help of a computer, Mandelbrot was the first to print out what is now called the Mandelbrot set for the case where z is a complex number. The Mandelbrot set is indeed a set of numbers, and his original monochrome printout showed in black those values of m for which the iteration did not diverge to infinity –

➤ A Quaternionic Julia Set. The Julia set is intimately linked to the Mandelbrot set, and here the iterations are performed using Hamilton's quaternions rather than complex numbers. Best viewed animated, this is a three-dimensional slice of a four-dimensional fractal.

that is, for which the iteration remained bounded. The incredible fine structure, with the characteristic jagged tendrils licking out like lightning, was gradually revealed as printers and computer graphics became more powerful. This simple system exhibited many of the characteristics which Mandelbrot strove to bring together. The self-similarity so characteristic of fractals can be seen by using a computer to zoom deeper into the set, where one finds mini-sets similar to the larger whole. Returning to the difference equation, for complex values of λ the iterations produce what Mandelbrot liked to call 'dragons'. The dreaded monsters of mathematical analysis had been reborn as beautiful creatures, and were welcomed into the family of mathematics.

The word 'chaos' is liable to misinterpretation, as in everyday language it is often synonymous with 'disorder'. But chaos theory is wholly deterministic: the Mandelbrot set will always look the same, and any starting value z will always lead to the same iterative sequence. The difference between a chaotic system and a random one is that randomness has no structure – it is the mathematical equivalent of white noise – whereas chaos does, albeit a very complicated and subtle one. However, although the generation of a fractal is a deterministic process, it is not predictable: there is no algorithm for deciding beforehand whether a point is inside the Mandelbrot set or not – the only way

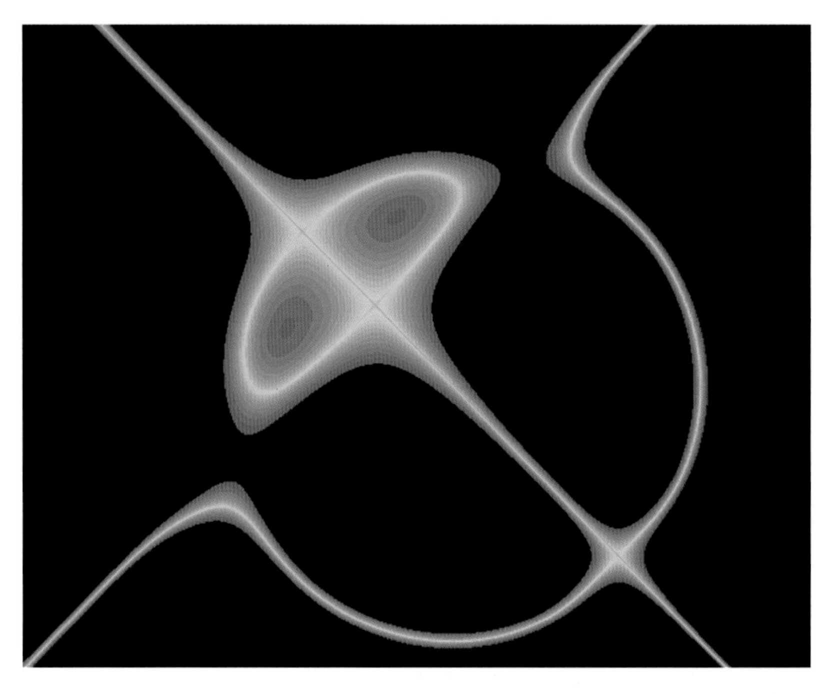

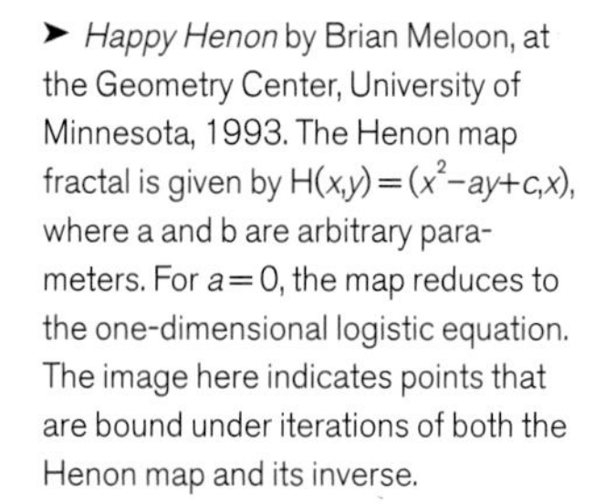

➤ *Happy Henon* by Brian Meloon, at the Geometry Center, University of Minnesota, 1993. The Henon map fractal is given by $H(x,y) = (x^2 - ay + c, x)$, where a and b are arbitrary parameters. For $a = 0$, the map reduces to the one-dimensional logistic equation. The image here indicates points that are bound under iterations of both the Henon map and its inverse.

to do this is to carry out the iteration. The coloured versions of fractals are a visual indication of how many iterative steps it takes for a point to tend to a very large value, and the intricate patterns with adjacent points of differing colours illustrate that such points will tend to diverge in the long run. Admittedly, a point on a computer screen is actually a pixel of finite size, but as the resolution increases so too do the intricacies. This is why chaotic systems are so difficult to predict. Although the iteration is deterministic, it is very sensitive to the initial value, and when it is used to model real-world systems the errors in taking initial measurements become magnified, The fact that many natural dynamical systems behave chaotically remains one of the mysteries of our universe.

Chaos theory may appear to give a fairly depressing view of the universe as a place of instability destined to dissipate under the relentless tyranny of the second law of thermodynamics. And yet the universe is full of structures, from the metronomic beat of pulsars to the exquisite convolutions in a DNA molecule. The onward march of entropy appears to be reversed, at least locally – the perfume has been put back into the bottle. The study of how such structures emerge is the province of what is termed complexity theory. The study of complex systems now includes a wide array of different areas, including chaos theory, artificial intelligence, emergent systems and automata.

Interest in complex systems arose in a number of different fields, and a key figure in bringing them together has been George A. Cowan. In 1942 Cowan, an expert in the chemistry of radioactive elements, was at the University of Chicago, where the Italian physicist Enrico Fermi was building the first atomic pile – the Allied forces feared that the Germans were already working on an atomic bomb. Fermi's early experiments and theoretical work were on the feasibility of initiating a chain reaction of sufficient energy to make a bomb. Cowan then worked on the Manhattan Project, and became Head of Research at the Los Alamos laboratory after the war. It was Cowan's group that analyzed the traces of Russia's first atomic explosion, he himself serving on the Bethe Panel, a covert group of scientists entrusted with monitoring Russia's nuclear capabilities, for some thirty years. During this period he became increasingly concerned with issues of science and public policy. He thought that traditional educational methods were not equipping scientists to see the broader connections between what appeared to be different disciplines, nor to see the links between science and the wider issues of politics, economics, the environment and morality. In 1982 Cowan stepped down from Los Alamos and took a seat on the White House Science Council. At the same time he sounded out the attitudes of colleagues to his dream of a centre dedicated to a more holistic study of all the quantitative sciences.

The growth of computing power was leading scientists to explore more complicated equations, not only ones with many more parameters, but also so-called non-linear equations. Much of mathematics up to this point had dealt with linear equations. This approach, though it had been very successful, was now beginning to seem a limitation

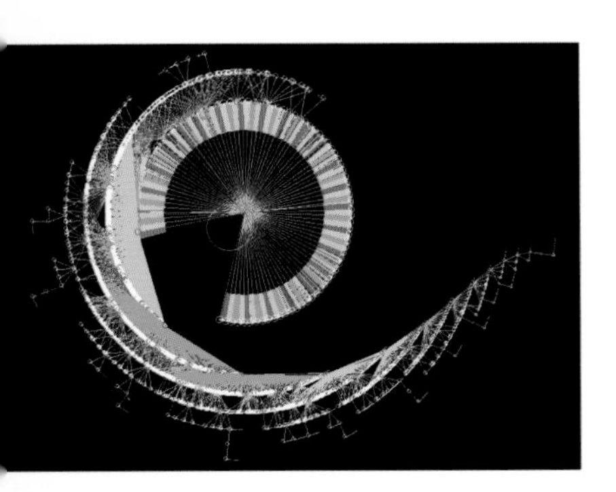

▲ Basin of Attraction - Ordered CA
Each node in this graph represents a snap-shot of the whole of a cellular automata (CA) universe, and is linked to the successor state in the next time-step of the evolving universe. This shows on ordered CA and the universe converges rapidly towards the basin of attraction.

when it came to accurately modelling more complex systems. But computers didn't care whether they were fed linear or non-linear equations – they just churned out numerical solutions at such speed, and in graphical form, that scientists and mathematicians now possessed what amounted to a new digital laboratory. It is with non-linear equations that we glimpse interconnections between variables previously treated as independent. Fruitful collaborations were already taking place between physicists and biologists, and Los Alamos even opened its own Center for Non-Linear Systems. But Los Alamos would not be diverted from its primary field of nuclear physics, so Cowan had to look elsewhere to build on the initial success at Los Alamos while expanding into other areas.

Cowan's colleagues were very receptive to the establishment of a new institute along the lines proposed, but at the time the sheer breadth of his vision made it difficult to define what this institute would actually do. The turning point came when Murray Gell-Mann joined the team. A leading theoretical physicist, it was Gell-Mann who had appropriated the word 'quark' from James Joyce's *Finnegans Wake* to label a new breed of subatomic particles. He was one of the leading advocates of a Grand Unified Theory to bring the fundamental forces of nature into a single, coherent framework. Now he wanted to go further, a Grand Unified Theory of *Everything*, from ancient civilizations to consciousness. He helped turn the widespread interest in the Institute into its actual founding. In 1984 the institute was incorporated as the Rio Grande Institute, as the preferred name, the Santa Fe Institute was then being used by a therapy service. By the end of the year the Institute had held its first workshops within the School of American Research in Santa Fe. Funding came from a variety of organizations, but as Cowan himself had actually made a small fortune in the 1960s setting up the Los Alamos National Bank, fund-raising was not an insurmountable obstacle. It became clear at this first meeting that some of the finest minds in their respective fields did have things to say to one another and shared many of the same concerns. These were essentially about emergent systems: the realization that wholes are greater than the sum of their parts, that from the interactions of many agents – be they particles, people, molecules or neurones – there emerges a complexity that is not apparent from the properties of the individual agents. Scientific reductionism seemed to work better going downwards, from complicated systems to simpler units, but was less successful at building upwards from such simple units to more complex structures. The initial excitement did not lead immediately to full funding, but the new centre did eventually get the Santa Fe name for itself. The first major backer came, appropriately enough, from the world of finance.

Banks and investment companies were becoming increasingly concerned about the ability of traditional economics to make accurate predictions about the evolution of the financial system. In 1987 the newly appointed chief executive officer of Citicorp took the opportunity to fund a workshop of economists and physicists under the auspices of the Institute. Certain physical systems had the same characteristics as social systems: they

▲ Basin of Attraction - Complex CA
This illustrates the evolution of a complex cellular automata (CA) universe. Convergence is much weaker and slower than in the ordered CA, and the graph shows characteristic branch formations. A chaotic CA graph would have thinner twig-like branches.

both exhibited similar mathematical relationships, and that mathematics was the mathematics of complex systems. In fact these are known as complex adaptive systems, exhibiting a number of negative and positive feedback mechanisms, and included systems such as the immune system, embryo development, ecologies, economic markets and political parties. The complexity arises from a mixture of competitive and cooperative tendencies, as such systems as these are continuously in a state of dynamic equilibrium, walking a tightrope between rigid order and chaos. And what was most astonishing was that such systems operated according to fairly simple rules, yet the intricate patterns somehow emerged from the interactions of these simple building blocks without there being a predetermined master template. Complexity was an emergent phenomenon. And the Santa Fe Institute was now firmly on the map.

One example of an agent as described above is an automaton. One 'world' of automata was known as the game of life, and was developed in 1970 by John Conway, a Cambridge mathematician. It was not so much a game as a miniature universe in which a two-dimensional grid was populated by evolving cells. Once the population was seeded, each cell would live or die depending on the number of live neighbouring cells – too many and it would die of overcrowding, too few and it would wither out of loneliness. Once set in motion, this universe would exhibit a wide range of structures, like oscillating diamonds, butterfly patterns and 'gliders' which seemed to walk across the landscape. John Von Neumann had started to investigate cellular automata way back in the 1940s, but his unfinished work was not edited and published until 1966, nearly ten years after his death. He had shown that there was at least one cellular automaton pattern which could reproduce itself – that reproduction was not the preserve of living organisms, and that software was independent of hardware, be it a computer or a brain. Francis Crick and James Watson's discovery in 1953 of the structure of DNA satisfied Von Neumann's analysis of the mathematical requirement of a self-reproducing system. In 1984 Stephen Wolfram pointed out that automata have deep similarities to non-linear dynamics. He categorized cellular automata into four 'universality classes'. Classes I and II produced static solutions within a short number of cycles, the first locking into a rigid, fixed structure, the second forming a periodic and stable one. Class III was a chaotic system showing no visible structure, while class IV included the game of life and other systems exhibiting emergent order. Christopher Langton further refined the classification and discovered that a system undergoing a phase shift, such as for example ice turning to water, would evolve from order to complexity to chaos. Cellular automata were heralded as a new life-form. Under the right conditions they could reproduce and could even act as a computer, not mimicking the hardware running the program, but acting as what Von Neumann and Turing would call a universal computer. The game of life demonstrated that life-like behaviour occurred in a state lying somewhere between order and chaos, a state of complexity within a finely tuned universe.

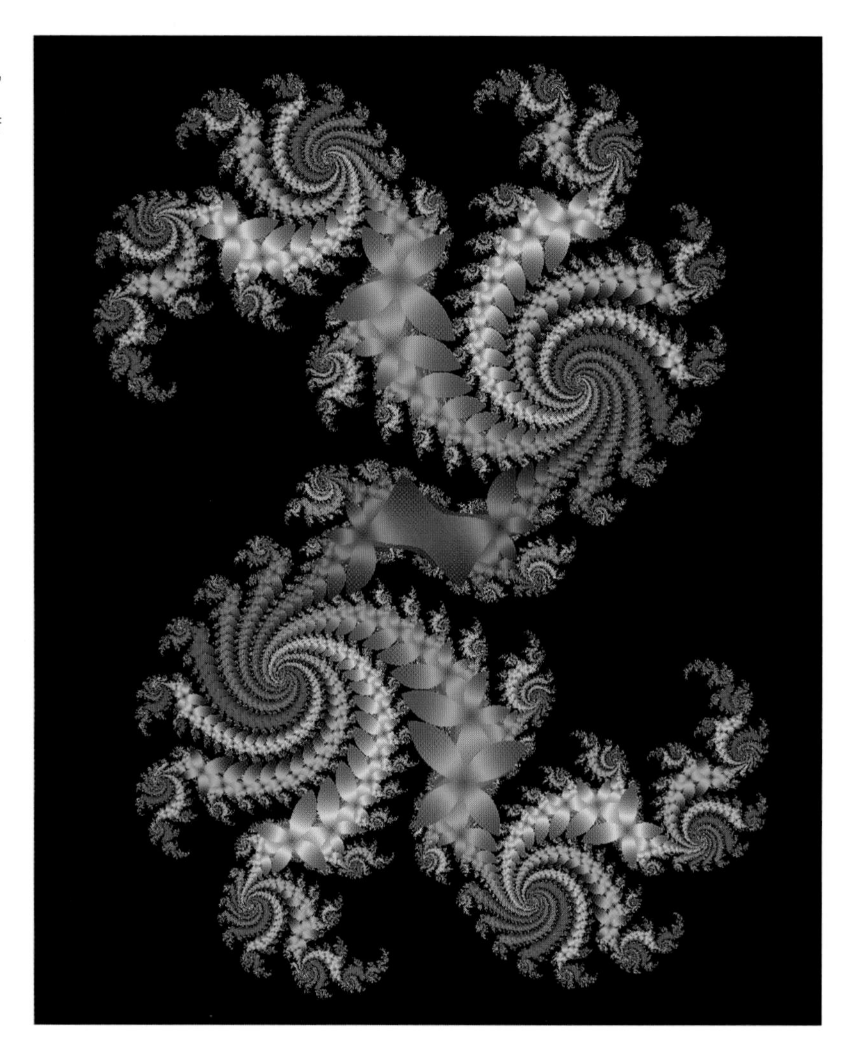

➤ This Mandelbrot dragon emerges from the generator function $f(z) = z^2 - m$, where z is a point in the complex plane and m the seed value. The black part of the plane shows the values of z for which the function tends to infinity as the number of iterations tend to infinity.

By 1990 the Santa Fe Institute had become the world's centre for research on complex systems. It is, perhaps, too soon to evaluate its impact, but what is clear is that the very nature of mathematics has changed, bringing with it a fundamental change in our philosophy of life and the structure of the universe. The Newtonian clockwork universe is dead, replaced by an evolutionary model of interconnected complexity. Mathematics continues to be as unpredictable as life.

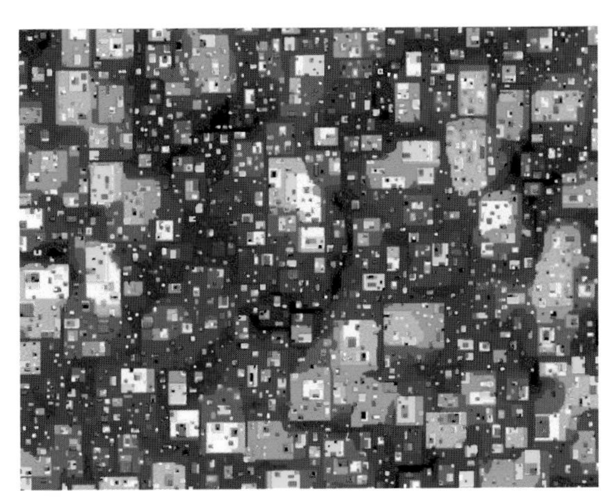

▲ Snapshots of two cellular automata universes. The colour of each pixel, or cell, denotes its state, which may change at the next time-step depending on the states of the cells around it. Starting from a randomly seeded universe, such basic rules can create evolving systems which manifest complex structures lying somewhere between order and chaos.

why is geometry often described as 'cold' and 'dry'? one reason lies in its inability to describe the shape of a cloud, a mountain, a coastline, or a tree. clouds are not spheres, mountains are not cones, coastlines are not circles, and a bark is not smooth, nor does lightning travel in a straight line.

More generally, I claim that many patterns of Nature are so irregular and fragmented, that, compared with Euclid ... Nature exhibits not simply a higher degree but an altogether different level of complexity. The number of distinct scales of length of natural patterns is for all practical purposes infinite.

The existence of these patterns challenges us to study those forms that Euclid leaves aside as being 'formless,' to investigate the morphology of the 'amorphous.' Mathematicians have disdained this challenge, however, and have increasingly chosen to flee from naturee by devising theories unrelated to anything we can see or feel.

I coined fractal from the Latin adjective fractus. The corresponding Latin verb frangere means 'to break:' to create irregular fragments. It is therefore sensible — and how appropriate for our needs! — that, in addition to 'fragmented' (as in fraction or refraction), fractus should also mean 'irregular,' both meanings being preserved in fragment.

scientists will (I am sure) be surprised and delighted to find that not a few shapes they had to call grainy, hydralike, in between, pimply, pocky, ramified, seaweedy, strange, tangled, tortuous, wiggly, wispy, wrinkled, and the like, can henceforth be approached in rigorous and vigorous quantitative fashion.

Benoît Mandelbrot, The Fractal Geometry of Nature, 1977

acknowledgements

Many thanks go to Professor Ivor Grattan-Guinness for his unwavering support for my various projects and for his patience and sound advice on the text. Any remaining errors are obviously my own. Thanks to Peter Tallack for his enthusiastic support of my vision for the book, and to Tim Whiting for nurturing it into shape, and to John Woodruff for those late night editing sessions. Thanks also to the Mathematics & Statistics Group at Middlesex University, and many of the members of the British Society for the History of Mathematics, as well as advice received from members of mailing lists. I would also like to thank friends who have kept the flame burning over the last couple of years: Eileen Barlex, Ben Dickey, Juri Gabriel, Peter Grecian, Dave Jackson, Chris Maslanka, David Prince, John Ronayne and David Singmaster. A lifetime's award to my parents, and apologies to anybody thoughtlessly left out.

select bibliography

There are many general histories of mathematics. The *Companion Encyclopedia of the History and Philosophy of the Mathematical Sciences*, edited by Ivor Grattan-Guinness, includes extensive bibliographies by chapter. More modestly priced histories include *The Fontana History of the Mathematical Sciences* by Ivor Grattan-Guinness, *A History of Mathematics* by Carl Boyer, *A History of Mathematics* by Florian Cajori, *A History of Mathematics, An Introduction* by Victor Katz and *Mathematical Thought from Ancient to Modern Times* and *Mathematics in Western Culture* by Morris Kline. Howard Eves has a series of anecdotal histories of mathematics, the most recent being *Return to Mathematical Circles.*

General introductions to ancient and non-European mathematics are *The Crest of the Peacock* by George Gheverghese Joseph, and *Science Awakening* by B.L. van der Waerden. On Greek mathematics there is *A History of Greek Mathematics* by Sir Thomas L. Heath, as well as his numerous translations including Euclid's *Elements.* For medieval mathematics, *Aristotle to Galileo* by A.C. Crombie (reprinted as *Medieval and Early Modern Science*) and *Physical Science in the Middle Ages* by Edward Grant, as well as his *A Source Book in Medieval Science*, and *The Beginnings of Western Science* by D.C. Lindberg. For Renaissance mathematics *The Invention of Infinity* by J.V. Field on perspective.

For the revolution in astronomy there are many original works now in translation, such as *The Harmony of the World* by Johannes Kepler and the *Two New Sciences* by Galileo Galilei. The most readable account of the story from Copernicus to Galileo is *The Sleepwalkers* by Arthur Koestler.

For a detailed overview of the calculus see *The Concepts of the Calculus* by Carl Boyer. There are many works on the scientific revolution. An insight into the social climate is *Ingenious Pursuits* by Lisa Jardine, also *The Revolution in Science 1500 - 1750* by A.R. Hall, and lively discussions in *On Giants' Shoulders* by Melvyn Bragg.

On cartography and astronomy, two readable and highly illustrated books are *The Image of the World* and *The Mapping of the Heavens*, both by Peter Whitfield. Chapters in his *Landmarks in Western Science* are also worthwhile.

A very readable history of early probability and statistics is *Games, Gods and Gambling* by F.N. David, and brought more up to date by *The Empire of Chance* edited by Gerd Gigerenzer.

On logic and transfinite numbers *The Principles of Mathematics* by Bertrand Russell is fairly accessible, *Georg Cantor: His Mathematics and Philosophy of the Infinite* by J. Dauben and *From Frege to Gödel: A Source Book on Mathematical Logic, 1879-1931*, edited by J. Van Heijenoort are also useful.

On computers, *The Computer, from Pascal to von Neumann* by Herman H. Goldstine includes many personal insights from participating in the early electronic computers, and *A Computer Perspective* by Charles Eames & Ray Eames is highly illustrated. On cryptography, *The Code Book* by Simon Singh.

On the visual representation of the mathematical sciences, both contemporary and historical, these are worth consulting: *The Fourth Dimension* and *Duchamp in Context* by Linda Dalrymple Henderson, *Science of Art* by Martin Kemp, *Artful Science* by B.M. Stafford, *Sacred Geometry* by R. Lawlor, *The Visual Mind* by Michele Emmer, *Gödel, Escher and Bach* by Douglas R. Hofstadter. There are also some conference proceedings, all available by searching the web, 'Bridges: Mathematical Connections in Art, Music and Science' (1998, 1999), 'NEXUS: Architecture and Mathematics' (1996, 1998), and 'ISAMA 99' (International Society of Art, Mathematics and Architecture).

On the mathematics of complex systems and chaos, there are many popular books which include discussions on these areas, from a historical point of view *The Fractal Geometry of Nature* by Benoit Mandelbrot and *Chaos* by J. Gleick, are worth reading, as are *Frontiers of Complexity* by Peter Coveney and Roger Highfield, and *Complexity* by M. Mitchell Waldrop.

There are a few collections of original mathematical writings, including *The History of Mathematics, A Reader* by John Fauvel and Jeremy Gray, *Classics of Mathematics* by Ronald Calinger, *The Treasury of Mathematics* by Henrietta Midonick, *A Sourcebook in Mathematics, 1200 - 1800*, by D.J. Struik, *On Mathematics and Mathematicians* by R.E. Moritz, *The World of Mathematics* by James R. Newman, and *Mathematical Maxims and Minims* by N. Rose.

And finally, a book which was an inspiration for the current work, popular in the 1960's and very much out of print, is *Mathematics in the Making* by Lancelot Hogben.

Websites worth checking out are the St Andrew's University 'History of Mathematics' website at
http://www-history.mcs.st-andrews.ac.uk/history/
with comprehensive links to other sites on the internet.
Eric Weisstein's 'World of Mathematics' is a very useful resource of modern mathematical terms and ideas with some historical references, found at http://mathworld.wolfram.com/
'Vismath' for visual mathematics at
http://members.tripod.com/vismath/
and the Geometry Center at
http://www.geom.umn.edu
and a Mathematics Museum based in Japan
http://mathmuse.sci.ibaraki.ac.jp/indexE.html
The British Society for the History of Mathematics homepage is at
http://www.dcs.warwick.ac.uk/bshm/

CDs focussing on visual mathematics are *Art and Mathematics* and *Life, the Universe and Mathematics*, both by Virtual Image. Also a CD *Escher Interactive* from Thames and Hudson Digital. The video, *VideoMath Festival at ICM98*, published by Springer, gives a broad range of modern work.

index

Page numbers in italics refer to illustrations

credits

THEARTARCHIVE 12, 13, 22, 39, 41, 43, 47, 48, 51, 55, 56, 57, 79, 93, 101, 102, 109, 112, 113, 114 (top right), 116, 118; AKG LONDON 45, 70, 85; ARTn 132; BRIDGEMAN ART LIBRARY 4 (top right), 9, 11, 15, 62, 63, 65, 76, 115, 169, 171; BRITISH LIBRARY 2, 25, 28, 29, 35, 49, 59, 61, 64, 67, 69, 86, 111, 139, 141, 159; ANDREW BURBANKS 149, 150, 179, 181; PAUL W. CARLSON (www.mbfractals.com) 187 (Image copyright Paul W. Carlson); CORDON ART 5, 125, 129 (All M.C.Esher works © Cordon Art, Holland); CHRISTIE'S IMAGES 32; DACS 165, Giacoma Balla, *Abstract Speed–The Car has Passed*, 1913 © DACS 2000, 169 Marcel Duchamp, *Nude Descending a Staircase*, no.2, 1912 © Succession Marcel Duchamp/ADAGP, Paris and DACS, London 2000, 171 Salvador Dali, *The Last Supper*, 1955 © Salvador Dali-Foundation Gala-Salvador Dali/DACS 2000; THE GEOMETRY CENTER 182; DAVID GRIFFEATH (http://psoup.math.wisc.edu/) 186 (Image created by David Griffeath); HULTON GETTY 18, 21, 27, 87, 89; THE MATHEMATICAL GUIDEBOOK: GRAPHICS By Michael Trott, Springer-Verlag, 2000 http//www.MathematicaGuideBook.com 130, 131; MUSEUM OF THE HISTORY OF SCIENCE 60, 90; NATIONAL GALLERY, LONDON 74; SCIENCE & SOCIETY 4 (bottom left), 17, 19, 53, 54, 58, 66, 73, 75, 77, 81, 83, 97, 99, 105, 106, 114 (bottom left), 117, 132 (bottom), 144, 145, 146, 173, 174, 175, 176, 178; KARL SIMS 162, 163; TATE GALLERY 165, 167; ANDY WUENSCHE/DISCRETE DYNAMICS LAB The images on pages 184 and 185 were produced from the website 'Complexity in Small Universes' at www.santafe.edu/~/Exh2/Exh3.html. For further information about DDLab and Wuensche's publications see www.ddlab.com or www.santafe.edu/~wuensch/ddlab.html

The image on p. 147 was created by the author using the software FRACTINT. To create your very own fractals and cellular automata, FRACTINT and WINFRACT are widely available as freeware on the internet.

copyright

First published in the United Kingdom in 2000 by Cassell & Co
This paperback edition first published in 2001 by Cassell Paperbacks

A CIP catalogue record for this book is available from the British Library
ISBN 1 841 88138 4

Design Director David Rowley
Editor Tim Whiting
Art Editor Austin Taylor
Designed by Mark Vernon-Jones
Printed and bound in Italy

Cassell & Co
Wellington House
125 Strand
London WC2R 0BB